Spatial Optimization
in Ecological Applications

Complexity in Ecological Systems

Complexity in Ecological Systems Series

Timothy F. H. Allen and David W. Roberts, Editors
Robert V. O'Neill, Adviser

Robert Rosen
Life Itself: A Comprehensive Inquiry Into the Nature,
Origin, and Fabrication of Life

Timothy F. H. Allen and Thomas W. Hoekstra
Toward a Unified Ecology

Robert E. Ulanowicz
Ecology, the Ascendent Perspective

John Hof and Michael Bevers
Spatial Optimization for Managed Ecosystems

David L. Peterson and V. Thomas Parker, Editors
Ecological Scale: Theory and Applications

Robert Rosen
Essays on Life Itself

Robert H. Gardner, W. Michael Kemp, Victor S. Kennedy,
and John E. Petersen, Editors
Scaling Relations in Experimental Ecology

S. R. Kerr and L. M. Dickie
The Biomass Spectrum: A Predator-Prey Theory of Aquatic Production

Spatial Optimization in Ecological Applications

John Hof and Michael Bevers

Columbia University Press
New York

Columbia University Press
Publishers Since 1893
New York   Chichester, West Sussex

Library of Congress Cataloging-in-Publication Data
Hof, John G.
  Spatial optimization in ecological applications / John Hof and Michael Bevers.
    p.   cm. — (Complexity in ecological systems)
  Includes bibliographical references (p.).
  ISBN 0-231-12544-5 (alk. paper) —ISBN 0-231-12545-3 (alk. paper)
    1. Spatial ecology—Mathematical models. I. Bevers, Michael.   II. Title.   III. Complexity
in ecological systems series.

QH541.15.S62 H65 2002
577'.01'5118—dc21   2001047717

Columbia University Press books are printed on permanent and durable acid-free paper.
Printed in the United States of America
c 10 9 8 7 6 5 4 3 2 1
p 10 9 8 7 6 5 4 3 2 1

To our parents: Gerrit, Ardyce, Guy, and Irene

CONTENTS

Preface   xi

1   Introduction   1
    Perspective   1
    Organization   3
    Methods   5
            Traditional Linear Programming Approaches   5
            Our Approach   8

PART I   Simple Proximity Relationships   11

2   Sedimentation   13
    Formulation   14
    Case Example   15
    Results   18
3   Stormflow Management   24
    Formulation   25
    Case Example   30
    Results   32
4   Natural Regeneration in Any-Aged Forest Management   42
    Formulation   44
    Case Example   46
    Results   47
            Solvability   47
            Solution Patterns   47
            Incomplete Initial Seeding   54
5   Combining Simulation with Optimization: Habitat Placement
    for the Northern Spotted Owl   59
    The Northern Spotted Owl   60

The Model   61
    The Connectivity (*f*) Function   65
    The Carrying Capacity (*g*) Function   66
Results   67
    Comparison of the Simulation Model with the Optimization Model   67
    Evaluation of the Plan   68

PART II   Reaction–Diffusion Models   73

6   Characteristics of the Discrete Reaction–Diffusion Model   78
Experiments   79
Populations Isolated in a Single Patch   80
    Effects of Patch Size   80
    Effects of Patch Shape   83
    Effects of Intrapatch Heterogeneity   83
Populations Occupying Multiple Patches   85
    Fragmentation of Contiguous Habitat   85
    Island Systems   89
    Effects of Patch Shape on Colonization   92
Discussion   93
7   The Basic Model: Habitat Placement for the Black-Footed Ferret   98
The Black-Footed Ferret   98
The Model   99
    Ferret Reintroduction in South Dakota   103
    Spatial Definition   104
    Ferret Dispersal   105
    Net Population Growth Rate   105
    Ferret Releases   106
    Ferret Carrying Capacity   106
Results   107
8   Population-Dependent Dispersal: Habitat Placement for the
Black-Tailed Prairie Dog   114
The Black-Tailed Prairie Dog   114
The Model   115
Results   120
9   Topography-Based Dispersal: Habitat Location for the Western
Prairie Fringed Orchid   125
The Western Prairie Fringed Orchid   126
The Model   128
    General Formulation   130

Specific Formulation 131
Landscape 134
Parameters 135
Results 136
Initial Seed Dispersal Assumptions 136
Conservative Seed Dispersal Assumptions 138
10 Habitat Edge Effects 142
Formulation 143
Case Example 147
Results 148
Forage Constraint Effects 149
Dynamic Scheduling 155
Management Scale Effects 159

PART III   Control Models 163

11 Strategies for Controlling Exotic Pests 167
Formulation 168
Case Example 171
Results 172
12 Strategies for Controlling Wildfire 183
Formulation 185
Case Example 187
Results 189
Extensions 196

PART IV   Using Optimization to Develop Hypotheses
About Ecosystems 201

13 Multiscaled Ecological Limiting Factors 203
Formulation 204
Case Example 205
Results 208
Reproduction and Dispersal Effects 210
Heterogeneous Habitat Effects 213
Discussion 216
Appendix 219
14 Carbon Fixation in Trees as an Optimization Process 221
Formulation 222
Case Example 226

	Results	228
	Basic Solution	228
	Sensitivity Analysis	229
	Discussion	231
15	Postscript	233
	References	235
	Index	251

PREFACE

When we completed *Spatial Optimization for Managed Ecosystems* (Hof and Bevers 1998), we immediately identified two shortcomings. First, although we suggested the use of optimization models for developing theoretical hypotheses about ecosystems, we presented no examples, so the point may have been lost. In this book, we include one section devoted to this topic and point to theoretical hypotheses throughout. Second, and more important, we feared that some readers of our last book might conclude that capturing ecological spatial relationships in optimization models requires the use of esoteric integer and nonlinear solution methods, implying that these relationships can be captured only heuristically or in small "toy models." In this book, we focus on capturing ecological relationships across a landscape with pragmatic optimization models that can be applied to real-world problems. We use linear programming primarily but also include two formulations for integer programming that are "integer-friendly." The model in chapter 14 is nonlinear but is still readily solvable.

Using linear programming makes it possible to include many thousands of choice variables and many thousands of constraints and still be confident of being able to solve problems with widely available software. To capture ecological relationships in linear programs, we think of the problem in terms of discrete difference equations, combined with the production system activity analysis applied in, for example, traditional timber harvest scheduling models. Even with this approach, we must often make simplifying assumptions about ecosystem function. The alternative approach would be to start with a more complex (e.g., nonlinear and nonconvex) model and investigate the use of heuristic procedures to approach optimized solutions (Boston 1999; Jager and Gross 2000). Just as it was difficult to choose between the X-15 and the Mercury program as the best approach for early space exploration (Von Braun et al. 1985), it is difficult to say which approach to optimizing ecological systems has more promise. Our emphasis here is on preserving optimality and exploring how much ecosystem function we can capture. We also emphasize solvability of large problems, including real-world case studies.

The methods applied are quantitative, and the book relies heavily on mathematical presentations. However, the reader can skip over the math and still get the general idea. Most of the chapters focus on examples. The book is written primarily for ecologists, resource economists, and management scientists. It is written for advanced undergraduate and graduate students and practicing scientists. As with our previous book, we hope that ecologists will forgive our many simplifying assumptions for the sake of gaining a different perspective on ecological problems.

This book has strong ties to our previous books (Hof 1993; Hof and Bevers 1998) but actually repeats little of the material found in them. It is put together largely from material we have published elsewhere in smaller pieces. To our coauthors of these previous works (identified with a footnote at the beginning of each chapter) we are gratefully indebted. We also thank the U.S. Department of Agriculture Forest Service, especially Fred Kaiser, Denver Burns, Tom Hoekstra, Marcia Patton-Mallory, John Toliver, and Brian Kent, for supporting this work. Many thanks to Dorothy Martinez and Penny Williams, who worked so hard to complete the word processing through many revisions; to Jill Heiner for computer programming support; to Tony Baltic for analytical and GIS support; to Scott Powell and Beth Galleher for additional GIS support; and to Joyce Van De Water and Mike Knowles for technical assistance with the figures.

Spatial Optimization
in Ecological Applications

1

INTRODUCTION

Perspective

Turner (1989) broadly defines landscape ecology as the study of the effect of landscape pattern on ecological processes. In this book, we present ideas and methods for taking ecological processes into account in optimizing landscape pattern through the strategic placement of management actions over time and space. Webster's *Third International Dictionary* defines "optimize" as "to make as perfect, effective, or functional as possible." Chiang (1974:244) refers to optimization as simply "the quest for the best." He notes that "the first order of business is to delineate an *objective function* in which the dependent variable represents the object of maximization or minimization and in which the set of independent variables indicates the objects whose magnitudes ... [we] ... can pick and choose. ... We shall therefore refer to the independent variables as *choice variables.* The essence of the optimization process is simply to find the set of values of the choice variables that will yield the desired extremum of the objective function" (1974:244). In unconstrained optimization problems, the choice variables are independent in the sense that the decision made regarding one variable does not impinge on the choices of the remaining variables. In constrained optimization, a set of constraints is included, each of which limits the value of some function of the choice variables to be less than or equal to, equal to, or greater than or equal to a specified constant. All the models in the chapters that follow represent constrained optimization problems.

Why would we want to optimize a landscape pattern? It is important to note that most of the work in this volume applies to managed ecosystems, where human impact is taken as a given and the problem centers around managing that impact. If a given level of human activity is adverse to the ecosystem, it makes sense to minimize its impact. Likewise, if we are in a position to help create positive impacts but with limited resources, then it seems reasonable to maximize that impact subject to the constraints implied by the limited resources. When a model is desired to predict impacts or consequences, simulation approaches are a logical choice. However, if it is desired to prescribe management activities, optimization approaches can implicitly evaluate huge numbers of options and allow tradeoff analyses that might otherwise be impossible. For more reading on the use of optimization in the general problems of multiple-resource management, see Hof (1993).

The most common technique for solving quantitatively defined constrained optimization problems is a set of methods called mathematical programming. This set includes but is not limited to linear programming, integer programming, and nonlinear programming. As applications of mathematical programming in natural resource management have evolved past commercial forestry problems, capturing ecological functions and relationships has been a central challenge. In meeting this challenge, many researchers have resorted to nonlinear and integer programming methods. In fact, in our previous book (Hof and Bevers 1998), *Spatial Optimization for Managed Ecosystems,* we use nonlinear and integer formulations in all but two chapters. However, these models are difficult to solve, thus limiting the size of the application and limiting the confidence that the analyst has in obtaining the best solution.

In this book, we explore formulations that capture highly nonlinear ecological effects with spatial linear programs that can be solved with simplex algorithms (and two "integer-friendly" linear mixed-integer programs that can be readily solved with branch-and-bound or heuristic methods). This makes it possible to include many thousands of choice variables and many thousands of constraints and still be confident of obtaining an optimal solution. The feat of capturing nonlinearities in linear programs is accomplished here with a variety of formulation methods, but they all boil down to discretizing the problem so that the difference equations relating one discrete time period to another or one discrete land area to another are linear (at least as first-order approximations).

With the heuristic methods available today (see Reeves 1993), it is possible to approximately solve large nonlinear and integer programs with a degree of suboptimality that, for any particular case, can be difficult to determine. Nonlinear programs can capture ecological relationships more precisely and more directly than the linear programs we develop in this book but often must be solved with an unknown level of suboptimality. This presents the analyst with a difficult choice, to paraphrase Reeves (1993), between obtaining a more exact solution of a more approximate model (as with linear programming) and obtaining a less exact solution of a more precise model (as with nonlinear programming). In this book, we pursue the former course, recognizing the legitimacy of both (see Haight and Monserud 1990a or Bettinger et al. 1997 for examples of the latter course). A practical factor that might tip the scale in our favor is that the heuristic methods for solving nonlinear programs tend to require sophisticated analysts capable of writing their own solution software, whereas linear programming solvers are widely available, are highly automated, and are simpler to operate.

All our models involve ecological processes that are not completely understood and are significantly affected by random events. This may make some of our simplifying assumptions a bit more palatable, but it also points out the importance of using our models (and others) in an adaptive management process (Walters 1986). In such a process, ecological behavior (including the response to management actions) is monitored, and the results are fed back into model revisions and additional analysis to generate adjustments in management strategy. Because our models are process oriented, they are conducive to use in this analytical role.

Organization

The book is organized into four parts: "Simple Proximity Relationships," "Reaction–Diffusion Models," "Control Problems," and "Using Optimization to Develop Hypotheses About Ecosystems." An introduction develops the basic concepts for each part. In part I, models that account for simple proximity relationships are discussed. In chapters 2 and 3, two related models are presented: a model that accounts for the spatial relationship between timbering activity and the sedimentation effects in nearby stream channels and a model that accounts for the spatial effect of vegetative manipulation on stormflow during severe precipitation events. In these chapters, the landscape is characterized as a watershed, with land areas defined by their runoff properties relative to stream channels. Chapter 4 treats individual trees as harvest choice variables and addresses mixed-age conditions, taking the spatial aspects of natural regeneration into account. The landscape is thus characterized by the areas occupied by mature trees. Chapter 5 uses a uniform grid of hexagonal cells to represent spatial structure and shows how simulation and optimization can be combined to model spatial (proximity) relationships for animals whose life history is too complex to capture directly in a linear programming model.

Part II presents linear programs based on the reaction–diffusion models in ecology that simultaneously capture population growth and dispersal over time and space. Chapter 6 explores the characteristics of the discrete reaction–diffusion model used in chapters 7–10 for optimization purposes. Chapter 7 discusses the basic model with an example that locates habitat for the black-footed ferret. This chapter is the only overlap with our previous book (Hof and Bevers 1998) and is used here as a point of departure. Chapter 8 presents a case study of black-tailed prairie dogs with a formulation that features population-dependent dispersal behavior. This formulation and the model results are compared with those in chapter 7. Whereas the ferret model in chapter 7 uses uniform square cells to define the landscape, chapter 8 uses irregular shapes to identify patches of potential habitat. Chapter 9 models an ephemeral plant, where multiple life stages and sensitivity to climate are featured in addition to the topography-based dispersal of seeds (one of the life stages). The landscape is structured according to topographic features (hummocks and swales) that define habitat and dispersal under different climate scenarios. Chapter 9 adds habitat edge effects to the reaction–diffusion model, contributing a definition of edge based on multiple habitat needs that is usable in a dynamic allocation model.

In part III the focus is control, contrasted with the preceding models, which try to maximize populations. In a mathematical programming sense (Luenberger 1984), our preceding models are also control (i.e., spatial control) models, but we use the term *control* here to emphasize that we are now trying to minimize rather than maximize a result. Chapter 11 shows how a linear programming model can be used to capture reaction–diffusion relationships when it is desired to minimize a population instead of maximize it, as one might want to do in trying to control an invading exotic species. Chapters 11 and 12 both use uniform square cells to define the landscape, but chapter 12 features a model formulation that tracks timing of fire spread through the landscape cells as opposed to using discrete time periods with associated diffusion distances.

In part IV we demonstrate how ecological theory and empirical investigations might be enhanced through the development of refutable hypotheses with optimization models. In chapter 13, hypotheses regarding the impacts on populations of multiple limiting factors operating at different scales are developed with a linear program, treating the population as a long-term optimizer and using linear programming to find equilibria. In chapter 14, optimization analysis is applied to models of carbon fixation in trees, treating the organism as an optimizer with several different behavioral assertions considered. The hypotheses that result are not empirically tested but are demonstrated in numerical examples.

The central purpose of the book is to describe case studies and pragmatic examples. Chapters 5 and 7–9 are case studies and have sections that describe the specific case, the model, and the results. Chapters 2–4 and 10–14 describe pragmatic examples and contain formulation, case example, and results sections. Chapters 2–4 focus on forest management practices (and mitigating their ecological effects), and all use a similar forest management component in their models. Chapters 2 and 3 are closely tied together, as are chapters 7 and 8, with similar study areas and closely related problem definitions. Reaction–diffusion formulations are used not only throughout part II, but also in chapters 11 and 13 (in their respective contexts). Throughout the book, six different approaches for modeling fauna are demonstrated, six different approaches for modeling flora are demonstrated, six different methods of characterizing the landscape are demonstrated, and two different methods of handling dynamics are demonstrated.

Methods

Many improved solution algorithms have become available in recent years for solving integer and nonlinear programs (see Reeves 1993), but we maintain that the simplex algorithm still is one of the most powerful tools in management science. It can solve huge linear problems (on the order of 50,000 constraints and 100,000 choice variables, depending on model structure) and reliably obtain an optimal solution in reasonable computing time. Simplex solvers are widely available, as are very powerful matrix generators that build linear programs efficiently. To take advantage of this capability, however, it must be possible to formulate the problem within the proportionality and additivity assumptions of linear programming. This means that the research challenge in this book is to formulate the problems in the first place: If linear approximation is possible and useful, solution is routine. For context, we quickly review the traditional approach to natural resource allocation with linear programming. We then begin to develop our ecological approach to landscape-level optimization.

Traditional Linear Programming Approaches

The basic structure of the linear programs historically used to analyze managed natural resource planning problems is depicted in table 1.1. For simplicity, the example in table 1.1 includes only a single discrete time period and ignores many constraints, such as budget limitations and minimum output levels. Also, table 1.1 only includes three outputs: timber, recreation, and forage.

In table 1.1, the major column headings are types of land and resource

TABLE 1.1
A Simple Depiction of Traditional Linear Programs Used in Multiple-Use Forest Resource Management and Planning

	Land Type I		Land Type II			Resource Products			Constraint Type	Right-Hand Side
Variables:	$X_{1,1}$	$X_{1,2}$	$X_{2,1}$	$X_{2,2}$	$X_{2,3}$	P_1	P_2	P_3		
Constraints										
Timber	$A_{1,1,1}$	$A_{1,1,2}$	$A_{1,2,1}$	$A_{1,2,2}$	$A_{1,2,3}$	-1			=	$K_1 = 0$
Recreation	$A_{2,1,1}$	$A_{2,1,2}$	$A_{2,2,1}$	$A_{2,2,2}$	$A_{2,2,3}$		-1		=	$K_2 = 0$
Forage	$A_{3,1,1}$	$A_{3,1,2}$	$A_{3,2,1}$	$A_{3,2,2}$	$A_{3,2,3}$			-1	=	$K_3 = 0$
Type I	1	1							=	L_1
Type II			1	1	1				=	L_2
Objective function										
Net benefit	$-C_{1,1}$	$-C_{1,2}$	$-C_{2,1}$	$-C_{2,2}$	$-C_{2,3}$	B_1	B_2	B_3		

products. The $X_{1,1}$ through $X_{2,3}$ columns represent choice variables for the number of hectares allocated to alternative management prescriptions that could be applied in type I ($X_{1,1}$ and $X_{1,2}$) and type II ($X_{2,1}, X_{2,2}, X_{2,3}$) land. The timber, recreation, and forage rows (equations) in the matrix represent the resource flows that result from implementation of the management prescriptions. For example, $A_{1,1,1}$ is the output of timber for each hectare of land type I on which management prescription $X_{1,1}$ is implemented. The type I and type II rows are the land inputs to this production system. L_1 hectares of type I land are available, and L_2 hectares of type II land are available.

The products (P_1, P_2, P_3) are accounting columns (variables) that collect the outputs described in the first three rows (equations) into aggregate outputs for the area being analyzed. K_1, K_2, and K_3 are set at zero to force all product output levels into P_1, P_2, and P_3. The coefficients in the last row, the net benefits equation, describe the cost if 1 unit of X_{hj} is applied on land type h and the benefit if 1 unit of P_i is produced. Thus, for example, $C_{1,1}$ is the cost of prescription $X_{1,1}$ on 1 hectare, and B_1 is the benefit derived from 1 unit of timber output (P_1). This row is the objective function to be maximized in this example.

An algebraic representation of the model in table 1.1 would be as follows:

Maximize

$$\sum_{i=1}^{3} B_i P_i - \sum_{h=1}^{2} \sum_{j=1}^{J_h} C_{hj} X_{hj},$$

subject to

$$\sum_{h=1}^{2} \sum_{j=1}^{J_h} A_{ihj} X_{hj} - P_i = 0 \qquad i = 1, \ldots, 3$$

(i.e., for all i, or $\forall i$)

$$\sum_{j=1}^{J_h} X_{hj} = L_h \quad \forall h,$$

where J_h (the number of prescriptions) is 2 for type I ($h = 1$) and 3 for type II ($h = 2$) lands. Lower bounds of zero are implicitly assumed for all variables in most linear programming problems, a convention we adopt throughout this book. If scheduling for, say, four time periods is to be included, then the model would be modified as follows:

Maximize

$$\sum_{t=1}^{4}\sum_{i=1}^{3} B_{it} P_{it} - \sum_{h=1}^{2}\sum_{j=1}^{J_h} C_{hj} X_{hj},$$

subject to

$$\sum_{h=1}^{2}\sum_{j=1}^{J_h} A_{ihjt} X_{hj} - P_{it} = 0 \quad \forall i, t$$

$$\sum_{j=1}^{J_h} X_{hj} = L_h \quad \forall h,$$

where

X_{hj} = the number of hectares allocated to the jth management schedule (i.e., a prescribed schedule of treatments over time) for land type h,

A_{ihjt} = the amount of the ith output in the tth time period that results from 1 hectare being allocated to the jth management prescription for land type h,

P_{it} = the total amount of the ith output produced in time period t,

B_{it} = the discounted benefit per unit of P_{it},

C_{hj} = the discounted cost per hectare (over all four time periods) of the jth management schedule for land type h.

When the land units are given a well-defined spatial context rather than just being defined collectively (as all type I lands, for example), such models provide a starting point for spatial optimization. The obvious limitation in this approach is that many ecological relationships are not captured, especially with regard to spatial relationships over time and across the landscape. Capturing these relationships in linear programs is the challenge set forth for this book.

Our Approach

The model just described is typical of activity analysis approaches, where the choice variables are activities that create a vector of outcomes. Our methods certainly include activity analyses but also involve functional (usually mechanistic) relationships between the activity choice variables, between the state variables (described later) in different time periods and land units, or between the choice variables in one time period or land unit and the state variables in another time period or land unit. Because our variables represent segments or pieces of larger variables across continuous time and space, the linear relationships between our variables capture discrete, piecewise approximations of nonlinear functions with regard to the larger variables.

A simple example of this would be a state variable for potential population in land unit i and time period t (S_{it}) being a function of the management prescriptions applied to the other land units:

$$S_{it} = \sum_{j \neq i} d_{ijt} X_j.$$

The possible solution values to the vector of management prescription variables (X_j) actually represent a continuum of actions over time and space, but the d_{ijt} coefficient is a linear approximation that applies only to specific segments of those larger variables. The d_{ijt} coefficient varies between land unit pairs and scheduled management activity in this approximation. This simple proximity relationship is the foundation of part I.

In terms of dynamic relationships between state variables, let us start with a simple exponential population growth function:

$$N_t = N_0 e^{rt},$$

where N_t is the population at time t, N_0 is the initial population ($t = 0$), and r is the per capita rate of increase. The first derivative is

$$\frac{dN_t}{dt} = r N_0 e^{rt} = r N_t.$$

If we define discrete time periods, then the difference equation equivalent is

$$N_t - N_{(t-1)} = r N_{(t-1)}$$

or

$$N_t = (l + r) N_{(t-1)},$$

which is linear.

If we add a spatial element with J discrete land units, and a state variable in unit i (S_{it}) is growing but is also related to diffusion from state variables in the other land units (indexed by j), then the difference equation might look more like

$$S_{it} = \sum_{j=1}^{J} g_{ij}(1+r)\, S_{j(t-1)},$$

where the summation includes the case $i = j$. The g coefficient captures the relationship between the local populations for every pair of land units during the discrete time period. Notice that the g coefficient can precisely capture a linear relationship between any two land units during the discrete time period, or it may serve as a first-order approximation over that discrete time period for a wide variety of nonlinear relationships. This dynamic spatial relationship is the foundation of part II and is discussed further in its introduction.

Discrete time periods and land units are included in table 1.1, but the relationships between time periods and between land units that we will formulate would be very difficult to visualize in such a tabular form. We thus rely on the algebraic form and also rely on a visualization of the problem based on difference equations that goes beyond simple activity analysis. Undergraduate instruction in natural resource management typically relies on the tabular representation of linear programming, which may have limited the types of formulations that practicing resource analysts have visualized. Many modern matrix generators (e.g., Brooke et al. 1992) are oriented toward the algebraic representation, so model construction with our formulations is well supported with commercially available software.

PART I

SIMPLE PROXIMITY RELATIONSHIPS

This first part of the book is devoted to capturing simple, static ecological relationships that are based on how close different land areas or landscape features are to each other. We refer to these as "proximity" relationships, which should be distinguished from the "adjacency" relationships discussed throughout the forestry literature. Concerns about adjacent timber harvests emerged from legal and regulatory restrictions on the effective size of harvests; for example, if two seemingly legal-size clearcuts occur within a short time of each other in adjacent areas, the effective size of the clearcut may not be legal. Because the original motivations for the size limits included concerns for aesthetics, wildlife habitat, water quality, and so forth, addressing adjacency restrictions often has been interpreted as addressing those underlying concerns. A fundamental thesis of this book is that addressing the actual spatial issues requires capturing the specific spatial relationships or processes involved and that adjacency constraints are adequate only when the problem is undesired adjacency of management actions per se. Avoiding adjacency of management actions may actually be counterproductive for connecting wildlife habitat patches, creating or limiting edge, controlling the spread of exotic pests, or managing water flows. The best way to address such concerns is to capture them directly in the analysis. In attempting to do so in this book, we often must make specific (sometimes simplistic) assumptions about how ecosystems function, but we believe that this is a more productive path than meeting adjacency criteria that address the problem indirectly and sometimes inconsistently. We compare our approaches with the adjacency restrictions in chapters 2 and 3.

Adjacency constraints invariably require integer formulations and create models that are difficult to solve. Ironically, the proximity relationships discussed in this chapter, which directly capture ecosystem relationships at least as first-order approximations, all lead to models that are solved readily: linear programs in chapters 2, 3, and 5. Even in chapter 4, where the definition of single-tree choice variables mandates an integer formulation, the resulting model is shown to be integer friendly.

Chapters 2 and 3 are interrelated in that they both focus on water flows in forest watersheds. They use similar case examples, and both are based on the effect of forest management and the resulting forest cover on nearby stream sections. The proximity relationships in chapter 4 are created by the limits to spatial seed dispersal in natural regeneration processes of an any-aged forest. In chapter 5, we present our first model that analyzes wildlife habitat connectivity, but this chapter combines simulation methods with optimization methods to reduce connectivity to a proximity relationship. Dynamic models that analyze habitat connectivity by tracking animal dispersal processes are discussed in part II.

2

SEDIMENTATION

One environmental impact often associated with timber harvesting is an increased level of sedimentation in nearby water courses just after the harvesting operation and during the subsequent "green-up" period of forest regeneration and growth. Linear programming (and other mathematical programming) models have become standard tools in optimizing the schedule of timber harvests, on the basis of economic returns as well as other objective functions. The typical approach to accounting for sediment effects in these types of models (as early as Bottoms and Bartlett 1975 or Dane et al. 1977 and as recently as Bettinger et al. 1998) is to estimate the impact of every potential harvest at some downstream point in or below the watershed being managed without regard for the effects throughout the watershed. If only the sediment exiting the area being managed is of concern, then the traditional approach is reasonable. On the other hand, if the sediment levels in reaches throughout the managed watershed are of concern, then a more complicated spatial problem is involved.

In this chapter we develop a simple model formulation that focuses on the spatial relationships between timber management variables and sediment levels in water runoff courses throughout the watershed being managed over time. This allows sediment standards to be set anywhere in the watershed. We begin by presenting this formulation and then explore it with a simple case example.

This chapter was adapted from J. Hof and M. Bevers, Optimal timber harvest scheduling with spatially defined sediment objectives, *Canadian Journal of Forest Research* 30 (2000): 1494–1500, with permission from the publisher, the National Research Council of Canada Research Press.

Formulation

We define the land units as areas that drain into stream sections as a part of a watershed system. We define the stream sections so that their midpoints are the locations where sediment is potentially regulated and monitored. Typically, the stream sections are delineated by forks, confluences, and other drainage features that create natural end points. We use discrete time periods and a "Model I" formulation (Johnson and Scheurman 1977) for timber harvest scheduling. Timber harvests are then chosen so as to meet prespecified timber targets and minimize sediment objectives that are spatially defined. Sediment levels could also be constrained, while maximizing a timber or economic objective function. We use the following formulation:

Minimize

$$\sum_{j \in \theta} \sum_t S_{jt}, \tag{2.1}$$

subject to

$$\sum_i \sum_{m=1}^{M_i} v_{imt} Y_{im} = Q_t \quad \forall t \tag{2.2}$$

$$S_{jt} = \sum_{i \in \Omega_j} \sum_{m=1}^{M_i} b_{ijmt} Y_{im} \quad \forall j, \, t \tag{2.3}$$

$$\sum_{m=1}^{M_i} Y_{im} = L_i \quad \forall i \tag{2.4}$$

$$Q_t \geq R_t \quad \forall t. \tag{2.5}$$

Indexes

i indexes land units,
j indexes stream sections,
m indexes management prescriptions,
t indexes time periods.

Parameters

M_i = the number of potential management prescriptions included for land unit i,

v_{imt} = the timber volume obtained per square kilometer in time period
   t by implementing management prescription m in land unit i,
b_{ijmt} = the sediment (metric tons) that passes through the midpoint of
   stream segment j during time period t as a result of 1 km^2 of land
   unit i being allocated to management prescription m,
θ = the set of stream sections included in the objective function,
Ω_j = the set of adjacent and upstream land units that potentially affect
   sediment levels in stream section j,
L_i = the area (square kilometers) of land in unit i,
R_t = the timber volume to be harvested in time period t.

Variables

Y_{im} = the amount of land (square kilometers) in land unit i allocated to
   management prescription m,
Q_t = the total volume harvested (cubic meters) in time period t,
S_{jt} = the average sediment level (in metric tons per year) at the midpoint of stream section j during time period t.

Equation (2.1) minimizes the total sediment level across all time periods and
the selected stream sections. Including different subsets of the stream sections
in the watershed is explored in the case example. Equation (2.2) accumulates
the volumes harvested in each time period into the Q_t variables. Equation (2.3)
accounts for the sediment load in each stream section, in each time period, as
affected by the harvesting of adjacent and upstream land units (through the management prescriptions included). Equation (2.3) assumes that the sediment
effects of harvesting different land units are additive. We therefore total these
sediment effects for each time period. Equation (2.4) limits the allocation of
each land unit's area (to management prescriptions) by its size. Equation (2.5)
includes the constraints on timber harvests in each time period.

As in many models, the mathematics in equations (2.1)–(2.5) are simple,
but the determination of the coefficients (particularly the b_{ijmt} parameters) is
complex. We discuss this process in the next section, where we develop a stylized case example.

Case Example

We assumed five time periods of 10 years each, with all management actions
taking place at the beginning of each time period. We also assumed that trees
become available for commercial harvest at age 30. For each cell, we included
nine prescription options numbered in order: cut in time period 1, cut in time

period 2, cut in time period 3, cut in time period 4, cut in time period 5, cut in time periods 1 and 4, cut in time periods 1 and 5, cut in time periods 2 and 5, and no harvest.

We assumed that artificial regeneration occurs immediately after harvest. For this case example, we used a yield function from Rose and Chen (1977) (converted to metric units):

$$V(t) = 20,790 \, (1 - e^{-.0301t})^{.6463} \, (1 - e^{-.0212t})^{1.243},$$

t = age in years,

$V(t)$ = volume per square kilometer at age $t \geq$ 30 years, in cubic meters.

Figure 2.1 shows a hypothetical watershed with 20.66 km^2 total area, which we digitized so that a geographic information system (GIS) could be used to calculate land areas (L_i) and distances between land unit centroids and stream section midpoints. To generate the sediment coefficients, we used the following function:

$$b_{ijmt} = 300 \left[1/ \left(2^{A_{imt}/10} \right) \right] \left[2/ \left(D_{ij} + 1 \right) \right], \qquad (2.6)$$

where

A_{imt} = the age (since harvest) of forest in land unit i in time period t under management prescription m,

D_{ij} = the distance between the centroid of land unit i and the midpoint of stream section j.

We used this function to calculate a b_{ijmt} for each i–j pair and for each time period t, where we calculated A_{imt} for each management prescription m. Note that each stream section j is potentially affected by harvesting on any adjacent or upstream land unit i. In an application, these b_{ijmt} coefficients probably would be best estimated with a simulation model (as in Bettinger et al. 1998) or with direct empirical data. We relied on the results of a classic field study (Moring 1975). In that study, the 75-ha Needle Branch watershed was completely clearcut. Sediment, measured approximately 1 km from the centroid of the clearcut, peaked at 221.4 metric tons per year the year after the harvest and diminished to 127.9 tons per year in the seventh year after harvest. Therefore we constructed equation (2.6) so that sediment transported from harvest areas declines (exponentially) by 50% every 10 years. We assumed that sediment will diminish in inverse proportion to distance and scaled equation (2.6)

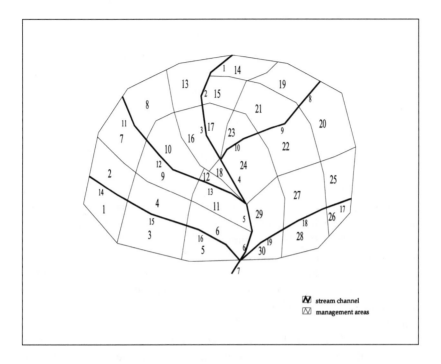

FIGURE 2.1

The case example watershed, with numbered stream sections
(small numbers) and management areas (large numbers).

so that at 1 km the sediment levels observed by Moring (1975) were approximated (221.4 tons per 75ha is approximately equivalent to 300 tons per
square kilometer). We calculated the A_{imt} coefficients in our matrix generator.
We used a GIS to calculate the D_{ij} coefficients and put them into the matrix
generator, which calculated the b_{ijmt} parameters for the linear program.

We began the analysis by maximizing the minimum time period's harvest
(across all five time periods) with no sediment constraints to determine the
maximum even flow for the five time periods (see Hof et al. 1986). That is,
we replaced equation (2.1) with the following:

Maximize

λ,

subject to

$\lambda \leq Q_t \quad \forall t,$

where λ is the minimum time period's harvest. This solution indicated a yield of 74,692 m^3 in each of the five time periods. We then set our timber targets at 50,000 m^3 (approximately two-thirds the maximum even flow) in all time periods.

Results

Figure 2.2 shows the harvest pattern for a model solution, with the sediment summed over all five time periods in stream section 7 (the reach just below the managed watershed) minimized. Table 2.1 presents the total sediment levels over time (from the objective function) in figures 2.2–2.4. In this solution, the model tends to avoid recutting young areas and cuts first in the areas least prone to causing sediment in stream section 7; as the remaining timber grows, the timber targets can be met with less area harvested. The areas harvested first are those that are farthest from stream section 7. As time goes on, the model selects areas closer to section 7, with increasing sediment effects despite the growth in timber volume per acre over time. This pattern seems inevitable if sediment effects are negatively related to the distance between the harvested area and the stream section of concern. We tried constraining the sediment levels (to 500 metric tons per year in section 7) and maximizing a discounted economic objective function (as in chapter 1), and the result was to harvest heaviest in time period 1, followed by reduced harvests with the same basic spatial pattern as seen in figure 2.2. With faster growth, an opti-

TABLE 2.1
Sediment Levels in the Solutions for Figures 2.2–2.4

	Figure 2.2	Figure 2.3	Figure 2.4
Minimized sediment in time period			
1	543.8	150.7	545.4
2	755.1	343.6	1,905.8
3	881.5	1,309.5	3,139.6
4	976.8	1,425.9	2,933.0
5	1,065.0	1,911.8	3,656.3
Total minimized sediment	4,222.2	5,141.5	12,180.1
Total maximized sediment	7,193.8	26,193.8	26,219.1

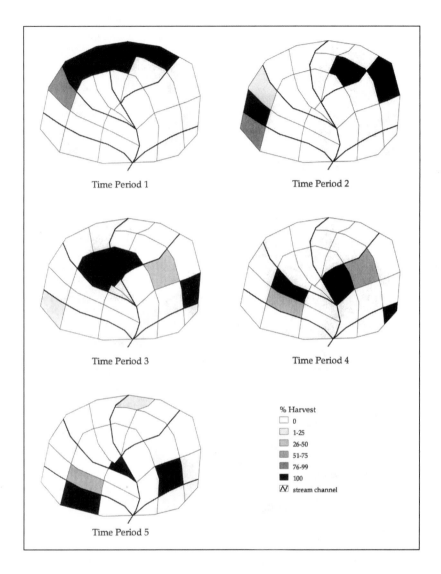

FIGURE 2.2

Harvest solution minimizing sediment in stream section 7, the watershed outlet.

mal solution might return to areas 7, 8, 13, 14, 15, and 19 in time period 4 or 5 rather than the areas shown in figure 2.2, but only if the growth is large enough to avoid large increases in area harvested to meet the timber target.

In figure 2.3, the sediment levels in stream sections 1–6 summed over all five time periods are minimized. This solution is very different from that in

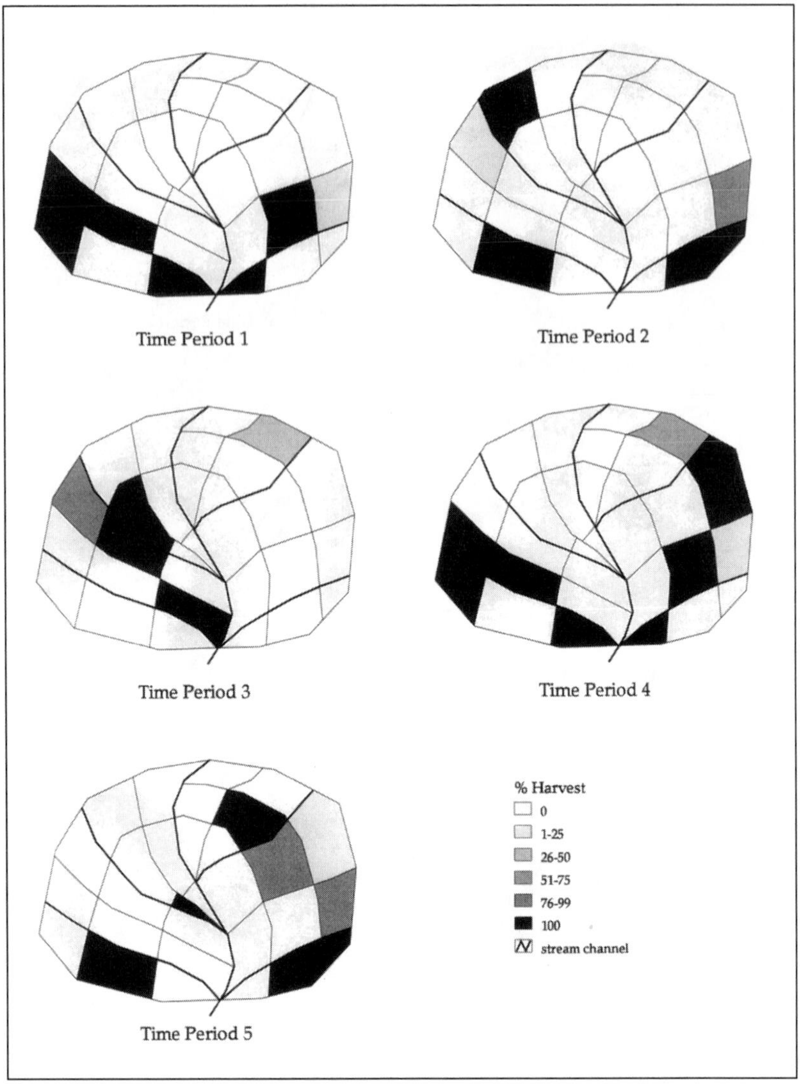

FIGURE 2.3

Harvest solution minimizing sediment in stream sections 1–6,
the main stream channel through the watershed.

figure 2.2. As one would expect, early harvests now avoid the upper reaches
of the main water course (sections 1–4) and are much closer to section 7. An
increasing sedimentation pattern is again displayed as more sediment-prone
areas (relative to sections 1–6) are harvested. The important point is that the

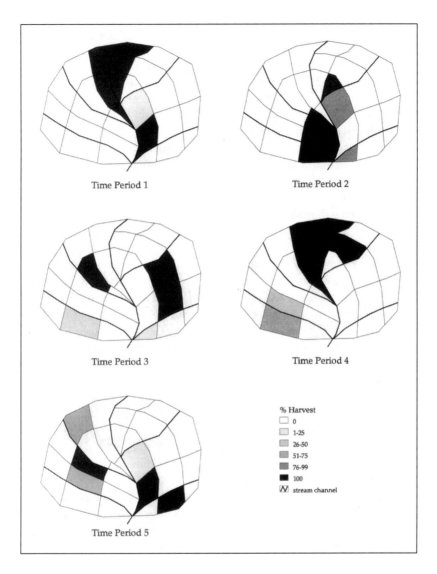

Time Period 1

Time Period 2

Time Period 3

Time Period 4

Time Period 5

% Harvest
0
1-25
26-50
51-75
76-99
100
stream channel

FIGURE 2.4

Harvest solution minimizing sediment in stream sections 8–19,
the upper reaches of the watershed.

spatial selection of the objective function strongly affects the spatial pattern
of solution.

This point is further demonstrated in figure 2.4, where the total sediment
in stream sections 8–19 over all five time periods is minimized. Here, the harvested areas tend to be close to sections 1–6 but with a preference for areas

13, 14, and 15 (more like figure 2.2) to the point that secondary harvests are taken in time period 4.

At least in our simple example, the spatial pattern of harvests is strongly affected by the spatial definition of the sediment objective. And the individual spatial relationships between management areas and stream sections are key in creating that sensitivity. As another experiment to look at the overall spatial sensitivity of our problem, we maximized the sediment objectives in each of the three models just discussed. The maximum sediment levels are given in the last line of table 2.1. Comparing the minimum and the maximum sediment levels for the three models indicates that the potential for spatial misallocation is greatest for the figure 2.3 model, which focuses on segments 1–6 (the maximum sediment level is five times the minimum level). This is interesting because the sediment-minimizing solution in figure 2.3 is the most different from the other two.

It is important not to generalize any results from our simple, stylized example. It demonstrates our modeling approach, however, and suggests that at least in some cases applying sediment objectives or constraints at the lowest (outlet) reach of a watershed may not be consistent with sediment objectives that apply throughout the watershed. Thus, the choice of spatial sediment objectives may be important, and we provide a simple method for including different spatial sediment objectives in linear timber harvest scheduling models.

A minimax approach (Luce and Raiffa 1957) might also be useful if a more even flow of sediment over time is desired. We should note that we have not accounted for stream temperature or timber transportation system effects. Some combination of the approach in Bettinger et al. (1998) and our spatial relationships within the optimization model might show promise along these lines.

Finally, in viewing the solutions in figures 2.2–2.4, it appears that in our case example sediment minimization tends to be associated with clustered harvests. Some authors (e.g., Roise 1990) suggest that adjacency constraints would be useful in managing water quality by limiting the effective size of clearcuts. There may be relationships between clearcut size and sediment discharge that our model does not account for, but our results suggest that in managing the spatial relationships between harvested areas and targeted stream sections, preventing adjacent harvests is not necessarily desirable. Much more evidence is needed for an empirical conclusion than we can provide here, but our results suggest an interesting hypothesis: Small, dispersed harvests may not necessarily be best in managing for spatially defined sediment objectives because concentration of areas that are isolated from targeted stream sections can take advantage of that spatial buffer.

This chapter presents the simplest formulation in the book, which amounts to spatially specifying where management takes place, indicating where the impacts are, and accounting for their proximity. Many problems might be addressed with this simple approach. The next chapter is similar but adds dynamic complexity in a nested schedule that simultaneously handles two time scales.

3

STORMFLOW MANAGEMENT

Our focus in this chapter is the effect of management activities such as timber harvesting and prescribed burning on storm event runoff (see Robichaud and Waldrop 1994). Physical soil and water effects such as channelization and sediment transport (and deposition) often can be attributed to the peak flow and total short-term water runoff from storm events (Janda et al. 1975). High peak flows are most commonly associated with storm damage, making confluences of stormflows at the same time and place particularly important. Because forest cover affects the magnitude and speed of storm runoff, the spatial arrangement of forest management activities can affect storm runoff impacts for many years. This chapter addresses the problem of spatially scheduling forest management activities over long periods of time, taking flow dynamics into account to mitigate the effects of targeted storms when they do occur.

In our previous book (Hof and Bevers 1998), we propose an integer programming approach to this general problem. This approach yielded models that were extremely difficult to solve, thus limiting its applicability. In this chapter we develop a linear programming formulation with the simplifying assumption that the most significant storm runoff affecting peak flows is from water that enters the channel system quickly. Therefore we assume that all storm runoff is accounted for in a detailed system of stream channels such that

This chapter was adapted from J. Hof and M. Bevers, A spatial linear program for optimally scheduling forest management to meet stormflow objectives, *Journal of the American Water Resources Association* 37, no. 3 (2001): 571–584, with permission from the publisher, the American Water Resources Association.

surface flow from one land unit to another is negligible. When this assumption is tenable, our linear programming approach can be applied to very large problems with little solution difficulty. We begin by presenting this linear programming formulation.

Formulation

Determining the location and timing of forest treatments to best meet vegetation and stormflow management objectives suggests that a spatial forest scheduling model be combined with a stormflow simulation and routing model. Combining two such models is not straightforward because time scales appropriate for modeling forest growth and management are substantially longer than the time scales appropriate for routing stormflows. Forest management often is modeled over a series of discrete planning periods that are years or decades long, whereas stormflows typically are simulated over a series of discrete time steps that may be only minutes or hours in length. Another difficulty is that over a 50- or 100-year forest management planning horizon, the timing of storm events cannot be predicted. In Hof and Bevers (1998), we address these challenges with a nested time schedule (see also Bevers et al. 1996). The approach is to select one or more targeted storm events for watershed planning purposes and model the stormflows from each storm over a sequence of appropriately short time steps within each forest management planning period. By setting upper bounds on the water outflow at selected locations for each storm event time step, one can schedule forest treatments over a long time horizon across the watershed to achieve stormflow objectives regardless of when the storm event occurs. Problems with spatially synchronized runoff from storms can then be managed with the long-term management actions that determine patterns of forest cover over time.

Initially, we will define a forest management objective to maximize total discounted net returns from timber harvests, subject to constraints on peak stormflows. For simplicity, we exclude forest values other than the direct returns from timber harvests. Using a watershed area subdivided into forest management units (microcatchments) whose boundaries define discrete stream channel sections (reaches), our linear programming formulation (for a single targeted storm event) is as follows:

Maximize

$$\sum_t p_t N_t - \sum_i \sum_{m=1}^{M_i} d_{im} Y_{im}, \qquad (3.1)$$

subject to

$$\sum_i \sum_{m=1}^{M_i} V_{imt} Y_{im} = N_t \qquad \forall t, \tag{3.2}$$

$$\sum_{m=1}^{M_i} Y_{im} = L_i \qquad \forall i, \tag{3.3}$$

$$W_{jqt} \leq \overline{W}_{jqt} \qquad \forall j, q, t, \tag{3.4}$$

$$W_{j1t} = K_{jt} \qquad \forall j, t, \tag{3.5}$$

$$W_{j(q+1)t} = (1 - c_j - b_j) W_{jqt} \tag{3.6}$$

$$+ \sum_{j' \in \theta_j} c_{j'} W_{j'qt}$$

$$+ \sum_{i \in \Omega_j} \sum_{m=1}^{M_i} g_{ijqtm} Y_{im}$$

$$+ Z_{jq} \quad \forall j, t$$

$$q = 1, \ldots, Q - 1,$$

$$a_{i1tm} = K_{itm} \qquad \forall i, t, m, \tag{3.7}$$

$$a_{i(q+1)tm} = (1 - r_{itm} - b_i) a_{iqtm} + u_{iqtm} \quad \forall i, t, m \tag{3.8}$$

$$q = 1, \ldots, Q - 1,$$

$$g_{ijqtm} = e_{ij} r_{itm} a_{iqtm} \qquad \forall i, j, t, m \tag{3.9}$$

$$q = 1, \ldots, Q,$$

with the following definitions:

Indexes

i indexes land units,

j and j' index stream sections,

m indexes management prescriptions (there are M_i of them for land unit i),

t indexes long-term planning time periods ($t = 1, \ldots, T$),
q indexes short-term time steps over the targeted storm event
   ($q = 1, \ldots, Q$).

Parameters

p_t = the discounted price per cubic meter of timber volume harvested
   in planning time period *t*,

d_{im} = the discounted cost per hectare of implementing management
   prescription *m* on land unit *i*,

V_{imt} = the timber volume per hectare obtained in planning time period
   t by implementing management prescription *m* on land unit *i*,

L_i = the area (hectares) in land unit *i*,

$\bar{W}_{jqt}$ = the maximum allowable stormflow in stream channel *j* during
   storm event time step *q* if it occurs in planning time period *t*,

K_{jt} = an initial water state base amount for stream channel *j* in planning time period *t*,

K_{itm} = the initial water state amount above long-term storage capacity
   per hectare of land in unit *i* that is allocated to management prescription *m* in planning time period *t*,

c_j = the proportion of water in stream channel section *j* (or *j'*) expected
   to flow downstream into the next stream channel section during
   each storm event time step,

b_j = the proportion of water lost from percolation, evaporation, or
   transpiration (often negligible during a storm event) from stream
   channel section *j* or per hectare in land unit *i* (b_i),

θ_j = the set of stream channel sections that contribute water directly to
   stream channel section *j*,

Ω_j = the set of land units adjacent to stream channel section *j* that contribute water directly to section *j*,

g_{ijqtm} = the water added to stream channel section *j* during storm event
   time step *q* if it occurs in planning time period *t* per hectare of land
   in unit *i* that is managed under management prescription *m*,

Z_{jq} = the exogenous water from direct precipitation and base flow
   expected to enter stream channel section *j* during storm event time
   step *q*,

a_{iqtm} = the water state amount above long-term storage capacity per
   hectare of land in unit *i* that is allocated to management prescription *m*
   during storm event time step *q* if it occurs in planning time period *t*,

r_{itm} = the proportion of water expected to flow out of land in unit *i* per
   hectare managed under management prescription *m* (with the resulting

forest cover conditions) during each storm event time step if it
occurs in planning time period t,

u_{iqtm} = the net amount of exogenous water from precipitation, snowpack, and other sources (after accounting for interception, snowmelt,
and similar processes) expected to enter each hectare of land in
unit i that is managed under management prescription m (with the
resulting forest cover conditions) during storm event time step q if
it occurs in planning time period t,

e_{ij} = the proportion of r_{itm} expected to flow into stream channel
section j.

Variables

Y_{im} = the number of hectares in land unit i allocated to management
prescription m,

N_t = the total volume harvested in planning time period t,

W_{jqt} = the water state amount in stream section j in storm event time
step q if it occurs in planning time period t.

Equation (3.1) is the economic objective function, which maximizes discounted net revenue. Equation (3.2) totals up harvested volume for each planning time period. Equation (3.3) limits the area that can be allocated to the
management prescriptions for each land unit to its size (this is an equality
because we include a no-harvest prescription). Equation (3.4) sets the upper
bounds on preselected stream sections for all planning time periods and storm
event time steps.

Equations (3.5)–(3.9) constitute the storm event simulator that is imbed-
ded in the linear program. Equation (3.5) initializes the water state variables
for the stream sections. Equation (3.6) determines the stream section water
states in subsequent storm event time steps with a difference equation. Then
downstream flows from each section and percolation, evaporation, and transpiration losses are deducted from the water state in the previous time step,
and flows from sections immediately upstream are added. Water runoff from
the land units adjacent to the stream section also is added, as is exogenous
water from direct precipitation and base flow entering the watershed. Stormflow sensitivity to forest cover (and thus to management activity) is captured
through the g coefficients on the management variables (Y) in equation (3.6).

Equations (3.7)–(3.9) involve only parameters and actually describe how
the g coefficients are calculated in our matrix generator. If the a parameters
were interactive variables (for example, because of surface runoff from one
land unit to another, as in Bevers et al. 1996), then the model's linearity would

be lost. Model linearity thus depends on land unit water states being predictable on the basis of the management activity in that land unit alone (and
not being affected by management activity on other land units). We anticipate
that this requirement can be met in many cases if the watershed drainage system of stream channels is modeled with adequate detail.

Equation (3.7) sets the initial water states for all the land units in all planning time periods. Equation (3.8) calculates the land unit water states in subsequent storm time steps with a lagged relationship. From the previous time
step's state, storm runoff and percolation, evaporation, and transpiration losses
are deducted, and the exogenous water from precipitation (with interception
netted out), snowpack melting, and any other source is added. Equation (3.9)
calculates the flows from each land unit into each stream section in each storm
time step (the g coefficients) on the basis of predetermined proportions. The
schedule of water added to each management unit from net precipitation,
snowmelt, and other sources is specifically modeled as a function of the storm
event time step as well as the planning period and selected forest treatment
schedule. Therefore this approach could be used to model rain-on-snowpack
storm events if it is assumed that snowpack levels and onsite snowmelt rates
(like our water states) are unaffected by other management units.

Stormwater comes from numerous sources, including surface runoff (Horton 1935), subsurface flows (Hursh and Brater 1944), and return flows (Musgrave and Holtan 1964), all of which appear as highly variable active source
areas during a storm event (Hewlett and Hibbert 1967; Troendle 1985). These
stormwater sources contribute to dynamic hillslope outflows into streams
(Roessel 1950). Consolidating stormwater from all sources on each forest
management unit into a single state variable with a proportional rate of outflow (which varies only by changes in forest cover) must be viewed as a first-
order approximation of a much more complex process. Nonetheless, with the
parameters presented in the following case study, a time series of proportional
outflows approximates a negative exponential process and does a reasonable
job of reproducing typical storm hydrographs with the inflow, storage, outflow, and summation characteristics described by Dunne and Leopold (1978).

No attempt is made here to simulate antecedent conditions involving water
deficits, such as soil moistures below field capacity. Such antecedent conditions ordinarily would not be of interest in problems aimed at limiting peak
stormflows. We assume that for the targeted storm event, initial land unit water
conditions are at least at long-term storage capacity, so all $K_{itm} \geq 0$. If water
is present in land unit i (at the beginning of the targeted storm event) above
long-term storage capacity, say from another storm event that occurred
recently enough that its water has not yet completely drained off land unit i,
then $K_{itm} > 0$, and this residual storm water would be added to (and modeled

in the same way as) the water from the targeted storm event. The amount of long-term storage capacity may be affected by land cover (age of forest since harvest), but this is irrelevant to our model. We will use the following case example to demonstrate and explore this model formulation.

Case Example

The timber harvest scheduling component of our case example is similar to the one in chapter 2. We assume five time periods of 10 years each, with all management actions taking place at the beginning of each time period. We also assume that trees become available for commercial harvest at age 30. For each cell, we include nine prescription options numbered in order: cut in time period 1, cut in time period 2, cut in time period 3, cut in time period 4, cut in time period 5, cut in time periods 1 and 4, cut in time periods 1 and 5, cut in time periods 2 and 5, and no harvest.

We assume that artificial regeneration occurs immediately after harvest. For this case example, we use a yield function from Rose and Chen (1977) (as in chapter 2, but converted to hectares):

$$V(t) = 207.9 \, (1 - e^{-.0301t})^{.6463}(1 - e^{-.0212t})^{1.243},$$
$$t = \text{age in years},$$
$$V(t) = \text{volume per ha at age } t \geq 30 \text{ years, in m}^3.$$

We use a price of \$17.66/m^3 of volume harvested, a cost of \$100/ha harvested, and a discount rate of 4%.

Figure 3.1 shows a hypothetical watershed like the one used in chapter 2 but rescaled to 744 ha total area, which we digitized so that a GIS could be used to calculate stream section lengths and land areas. The main stream channel includes sections 1–7 (with 7 being the exit stream section). The tributaries to this main channel are numbered 8–19. We set initial stream section water states (K_{jt}) at 4 m^3/m of stream section length for sections 1–7 and at 0 for sections 8–19. We assume that the targeted storm event occurs when land unit water is at long-term storage capacity (field capacity) and no residual storm water from any previous storm is present, thus all $K_{itm} = 0$.

We assume that water moves in the main stream channel at 300 m per 10-minute time step and at 340 m per 10-minute time step in the tributaries. This implies that

$c_j = 300 \div$ section length in meters for sections 1–7;
$c_j = 340 \div$ section length in meters for sections 8–19.

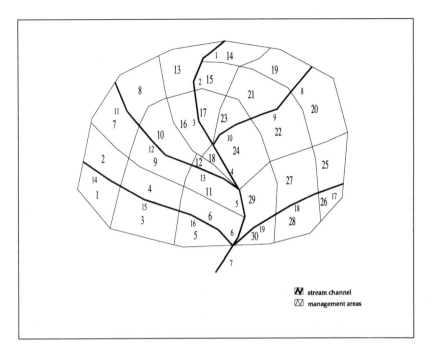

FIGURE 3.1

Case example watershed, with numbered stream sections
(small numbers) and management areas (large numbers).

The landscape must be scaled so that the c_j coefficients do not exceed 1. We also assume that $b_j = 0$ and $b_i = .001$, reflecting minimal percolation, evaporation, and transpiration during a storm event. Base flow (Z_{jq}) is set at 1200 m³ for $j = 1$ and all q and at 0 otherwise (treating direct precipitation into stream channels as negligible).

The rate of runoff (r_{itm}) from the land units into the stream sections is a function of stand age, as is the interception loss from precipitation. Values for these parameters are given in table 3.1. The precipitation (before interception loss) for the targeted storm event is assumed to be 2.12 cm in the first two storm event time steps (1.06 cm each). This translates into 106 m³ of water per hectare in each of the first two time steps. The θ_j and Ω_j sets were determined from the landscape map, and the e_{ij} proportions were set according to the spatial drainage characteristics of the watershed.

With water flowing at the speed of 300 m every 10 minutes (see calculation of c_j) and each stream section initially containing 4 m³/m (K_{jt}), 1200 m³ leaves sections 1–6 in each 10-minute time step and enters sections 2–7 in

TABLE 3.1

Land Unit Hydrologic Parameters by Forest Cover Age

| Age (yr) | Outflow Rate (r) | Interception Loss Rates | |
		q = 1	q = 2
0	.11	.1	.02
10	.09	.24	.088
20	.07	.38	.156
30	.05	.52	.224
40	.03	.66	.292
50+	.01	.8	.36

each 10-minute time step. With $Z_{1q} = 1200$ for all q, 1200 m^3 of base flow enters section 1 in each 10-minute time step. Thus, if all $a_{iqtm} = 0$, the main stream section is in equilibrium at the initial conditions, with the land units at field capacity (but with no additional water present) and the tributaries empty.

Results

To demonstrate the simulation portion of the model, we forced all management variables into a no-harvest alternative and solved the model with two different age structures: a mature (60 years old) forest and a completely cutover landscape. Figure 3.2 shows hydrographs for main stream channel sections 1, 3, 5, and 7. All hydrographs start at the level of 400 m^3 per 100 m of stream section length. The stormflow response in the downstream sections is significantly higher than in the upstream sections, and the water runoff is much faster and dramatic for the cutover area than the mature forest, as one would expect. Of particular note for interpreting our optimization results is that the cutover forest hydrograph peaks much earlier than the mature forest hydrograph, enhancing the tradeoffs that are possible over time and space.

Figure 3.3 shows the harvest solution when we maximized discounted net revenue, subject to a constraint that the water state in stream section 7 cannot exceed 7000 m^3 in any storm event time step in any planning time period. The initial forest age was set at 60 years in all land units for this and all subsequent solutions. The heaviest harvest takes place in time period 1 (40,975 m^3), followed by 9132 m^3, 11,412 m^3, 14,524 m^3, and 17,196 m^3 in time periods 2, 3, 4, and 5, respectively. The harvest in time period 1 is split between areas that are high upstream and low downstream. The area in between acts as a timing

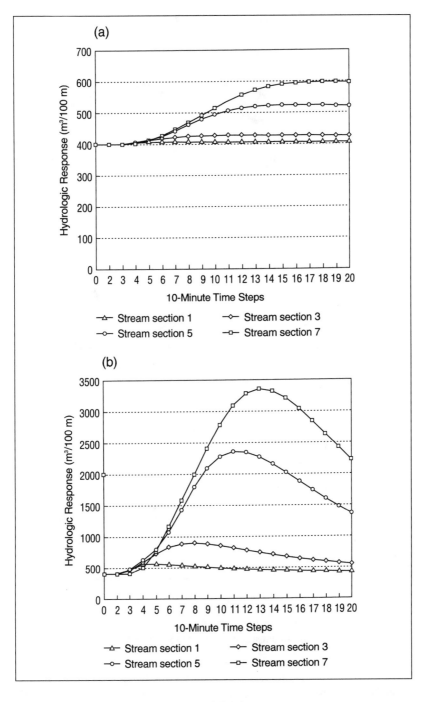

FIGURE 3.2

Modeled hydrologic response in cubic meters per 100 meters of stream section length, for **(a)** a watershed covered with 60-year-old forest and **(b)** a watershed following complete forest removal.

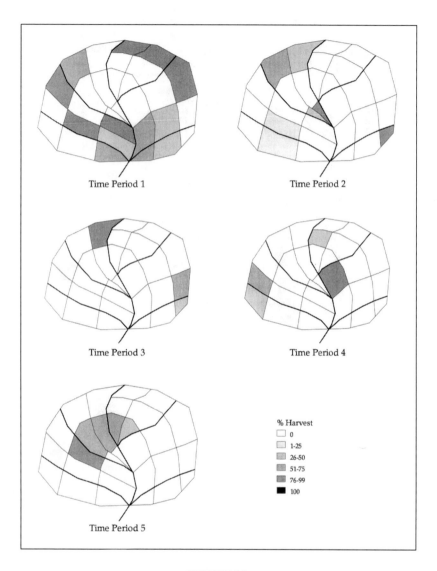

Time Period 1

Time Period 2

Time Period 3

Time Period 4

Time Period 5

% Harvest
☐ 0
☐ 1-25
▨ 26-50
▨ 51-75
▨ 76-99
■ 100

FIGURE 3.3

Harvest solution maximizing discounted net revenue, with stormflow
in stream section 7 limited to 7000 m³.

buffer (it is still mature forest) so that during the targeted storm event, the
water from the upstream harvested areas does not reach stream section 7 until
the impact on stream section 7 from the downstream harvests has at least partially run its course. In subsequent time periods, the rest of the upstream land
units and the middle area are harvested gradually. The discount rate encour-

ages the heavy harvest in time period 1, and it is feasible relative to the storm-flow constraint because the residual mature forest is available to act as a timing buffer. After that, only much smaller harvests are feasible (the stormflow constraint is binding in at least one storm event time step in all five planning time periods). Harvests do increase from time periods 2 through 5 as the area recovers from the initial harvest. A harvesting solution closer to even flow would result from initial conditions that have more forest ages represented, but this solution demonstrates the spatial sensitivity of the model.

To explore this spatial sensitivity further, we performed the following analysis. First, we maximized the minimum time period's harvest (across all five time periods), with no stormflow constraints (3.4), to determine the maximum even flow for those five time periods (see Hof et al. 1986) by replacing equation (3.1) with the following:

Maximize

$$\lambda,$$

subject to

$$\lambda \leq N_t \quad \forall t,$$

as in chapter 2. This solution indicated a timber yield of 26,889 m^3 in each of the five time periods. We then set timber constraints such that 13,500 m^3 (approximately half the maximum even flow) had to be harvested in each time period. Then we minimized the maximum (Luce and Raiffa 1957) water state in stream section 7 across all storm event time steps and all planning time periods using

Minimize

$$\lambda,$$

subject to

$$\lambda \geq W_{7qt} \quad \forall q$$
$$\forall t$$
$$N_t = 13,500 \quad \forall t$$

in place of (3.1) and (3.4). The harvest solution is shown in figure 3.4. This solution starts harvesting upstream in time period 1, gradually moving downstream through time period 4, and then splits between downstream and

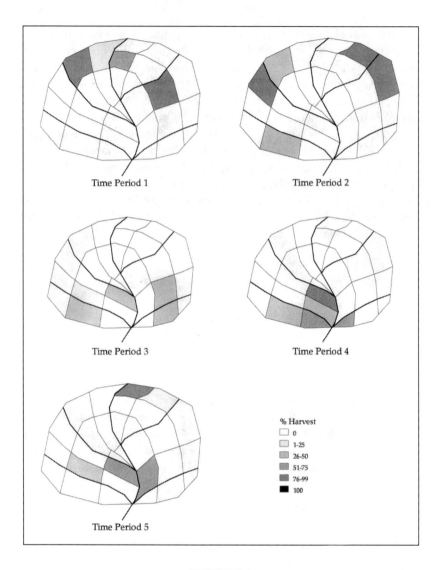

FIGURE 3.4

Harvest solution minimizing the maximum water state in stream section 7 across all time periods and time steps, with harvest fixed at 13,500 m^3 in each time period.

upstream harvests in time period 5. In this solution, time steps 14 and 15 in time period 3, time step 14 in time period 4, and time steps 13 and 14 in time period 5 are binding, at a level of 5,734 m^3. This strategy avoids the confluence of large stormflows in the same storm event time step in stream section 7, associated with the even flow constraint imposed. To explore spatial sensitivity, we next maximized the sum of the binding water states in this solution using

Maximize

$$W_{7,14,3} + W_{7,15,3} + W_{7,14,4} + W_{7,13,5} + W_{7,14,5},$$

subject to

$$N_t = 13,500 \quad \forall t$$

in place of (3.1) and (3.4). This harvest solution is shown in figure 3.5, which is something of a mirror image of that in figure 3.4. The average water level in the five variables whose sum was maximized is 6818 m^3, 19% higher than the minimized maximum. If just $W_{7,14,5}$ is maximized, it is 33% higher than the minimized maximum. For at least this simple example, the spatial and temporal arrangement of harvests could influence peak stormflow at the outlet stream section 7 substantially.

It is also possible that, similar to sediment levels in chapter 2, stormflow effects are of concern upstream in the watershed instead of just at the watershed outlet. To investigate this possibility, we repeated the previous analysis but focused on the upstream sections (8–19). We again set timber harvests at 13,500 m^3 in each planning time period and then minimized the maximum water state per meter of stream section length (because sections of different lengths are included) across all time steps, all time periods, and stream sections 8–19 using

Minimize

$$\lambda,$$

subject to

$$\lambda \geq W_{jqt}/S_j \quad \forall q$$
$$\forall t$$
$$j = 8, \ldots, 19$$
$$N_t = 13,500 \quad \forall t$$

in place of (3.1) and (3.4). This harvest solution is shown in figure 3.6. Clearly, in this example the spatial harvesting strategy is strongly affected by the spatial definition of the objective. The spatial strategy in figure 3.6 is basically to

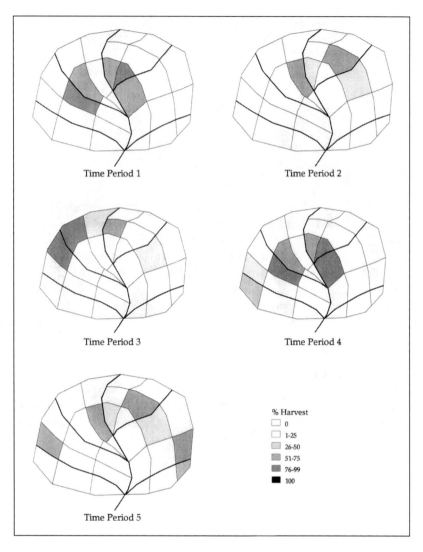

FIGURE 3.5

Harvest solution maximizing the binding stream water state variables
from figure 3.4, with harvest fixed at 13,500 m^3 in each time period.

disperse harvests, with some concentration upstream from the targeted tributaries and near the main stream channel (sections 1–3) in planning time periods 2 and 5. If these results are representative, then a strategy focused on the
outlet stream section may not be best for the upstream sections and vice versa.
In a given planning situation, spatial specification of the stormflow objective

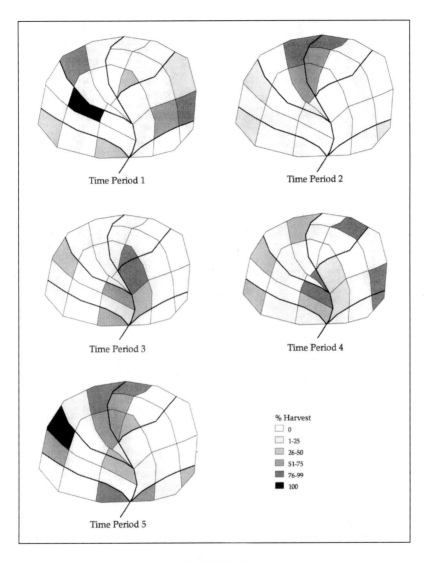

FIGURE 3.6

Harvest solution minimizing the maximum water state per meter of section length across stream sections 8–19, all time periods, and all time steps, with harvest fixed at 13,500 m^3 in each time period.

may not be obvious but could strongly affect the spatial strategy ultimately arrived at.

There were 24 W_{jqt} variables that were binding in the figure 3.6 solution involving stream sections 8, 10–14, 16, 18, and 19. We maximized the sum of

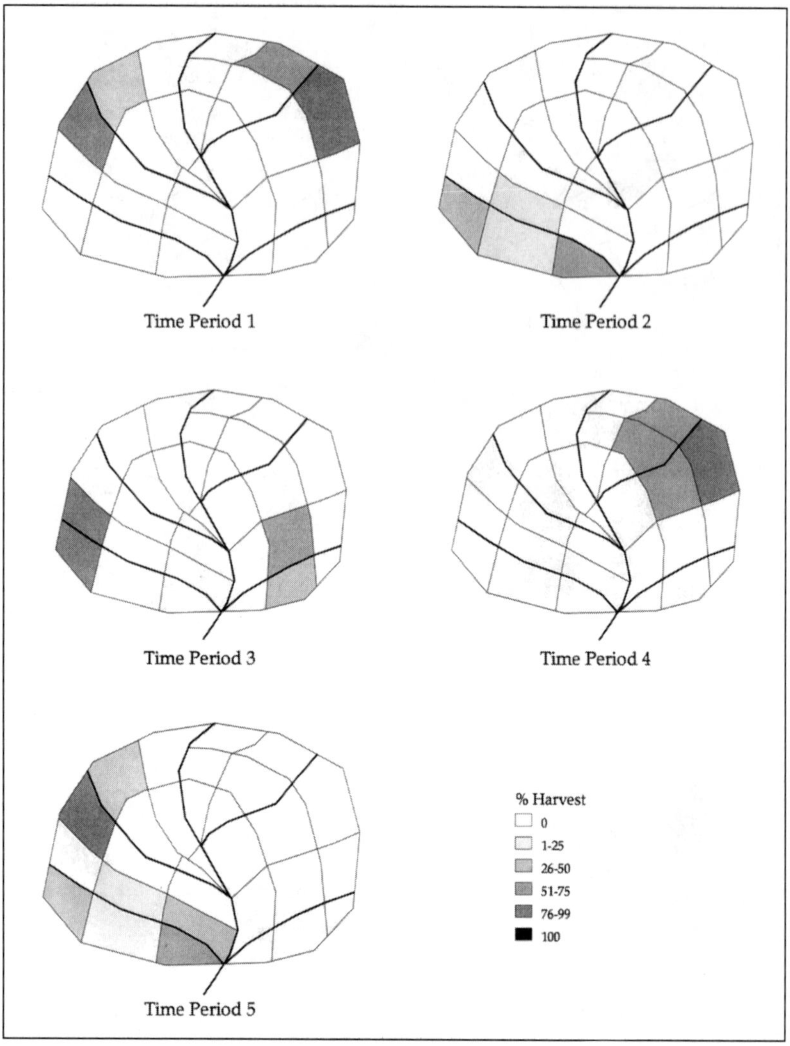

FIGURE 3.7
Harvest solution maximizing the binding stream water state variables
from figure 3.6, with harvest fixed at 13,500 m³ in each time period.

these, as we did for the solution in figure 3.5, with the same timber harvest
constraints set at 13,500 m³. The resulting harvest solution is shown in figure
3.7. Again, this solution is very different spatially from its counterpart in figure 3.6, being concentrated away from the main channel rather than being dif-
fused. The sum of the water state levels in the 24 binding W_{jqt} variables in the
figure 3.6 solution was 23,400 m³, whereas it is maximized at 46,524 m³ in

the solution shown in figure 3.7. For this example, the spatial and temporal placement of harvests could affect upstream peak stormflow levels by even more than it might affect peak flow levels at the outlet.

The primary contribution of this chapter is a linear programming model formulation for spatial and temporal allocation of forest management that limits peak stormflow effects. The use of nested time schedules provides a method for linking forest and stormflow management models and provides a useful mechanism for designing watershed decisions around one or more storm events (Bevers et al. 1996). With these constructs, it is possible (at least in principle) to optimize the spatial layout of forest management activities over time to create a forest cover schedule that limits undesirable effects from stormflow given the uncertainty about the timing of the storms themselves.

Our approach requires a number of simplifying assumptions, especially that the effect of surface runoff between land units on peak flows is negligible. In general, we assume that land unit outflow rates during storm events are strictly a function of watershed geomorphology and vegetative cover. Our spatially explicit hydrologic model is simple but resulted in reasonably behaved hydrographs while allowing linear solution methods.

As in chapter 2, it is important not to generalize any results from our example; it is included to demonstrate our formulation. In this simple example, peak stormflows were shown to be spatially sensitive to the temporal and spatial allocation of timber harvests, and solution strategies were shown to be sensitive to the spatial definition of stormflow objectives. Also note that our example solutions reflect a process-oriented model of stormflow rather than a prespecified prescription such as limiting harvest area size. As in chapter 2, there may be relationships between harvest size and peak stormflow that our model does not account for, but our results suggest that preventing adjacent harvests is not necessarily desirable. Again, much more evidence is needed for an empirical conclusion than we can provide here, but small, dispersed harvests may not necessarily be best in managing for spatially defined stormflow objectives. Some concentration of harvested areas may be able to take advantage of spatial timing buffers between them (see, for example, figure 3.3). Water quality and stormflow effects are complicated spatial problems, and further research is certainly needed beyond our work in this chapter and chapter 2. In chapter 4, the proximity relationships involve natural regeneration of trees, but the timber harvest scheduling component of that model is similar to the one used here (though adapted to single-tree choice variables).

4

NATURAL REGENERATION
IN ANY-AGED FOREST MANAGEMENT

The future of forest management in North America probably will involve less and less even-aged management (which leaves a residual forest with significant areas of homogeneous tree ages). Harvest systems other than clearcutting have been applied to a significant portion of the nation's forests for many years, and this portion can only be expected to increase given recent forest policy trends. Clearcutting is coming under increasing criticism on public lands; legislation has been proposed that would ban clearcutting on all public lands (Anonymous 1998b). On private land, policy trends also indicate reduced even-aged management in the future. For example, MacMillan Bloedel, Canada's largest forest products company, recently announced plans to implement partial retention management on even its most intensively managed lands (Anonymous 1998a). Weyerhauser has acquired MacMillan Bloedel and announced their intent to continue this new policy.

Many studies have formulated optimization models for uneven-aged forest management (e.g., Adams and Ek 1974; Buongiorno and Michie 1980; Haight et al. 1985; Michie 1985; Bare and Opalach 1988; Gove and Fairweather 1992; Anderson and Bare 1994). These studies assume a steady-state solution and view the stand-level uneven-aged management problem as one of determining the optimal diameter class distribution, the optimal species mix, the optimal cutting cycle length, and also an optimal conversion strategy for stands not initially in the desired steady state (Hann and Bare 1979; Gove

This chapter was adapted from J. Hof and M. Bevers, Optimizing forest stand management with natural regeneration and single-tree choice variables, *Forest Science* 46, no 2 (2000): 168–175, with permission from the publisher, the Society of American Foresters.

and Fairweather 1992). We will ignore the species mix problem and focus on a single species. The remaining problem characterization presumes that a single, fixed-diameter class distribution and a fixed cutting cycle are desired. These steady-state assumptions make the problem more tractable, with choice variables based on diameter classes and a cutting cycle that apply across the stand. Haight (1987) and Haight and Monserud (1990b) note that optimal forest management may not fit either the even- aged management constraints or these steady-state uneven-aged management restrictions, and they suggest "any-aged" management as an unfettered alternative.

In this chapter we adopt this unrestricted view of the problem. Our basic hypothesis is that the natural choice variable for any-aged management is the individual tree. The idea of using individual trees as choice variables is not new, but previous work has focused on optimizing with various single-tree growth models (e.g., Yoshimoto et al. 1990; Pukkala and Miina 1998). In contrast, our approach is to optimize forest stand management based on explicit spatial relationships between trees. We then explore whether optimal solutions at the stand level tend toward steady-state uneven-aged management, even-aged management, or some other pattern with these choice variables.

From a management science perspective, an individual-tree approach is daunting because of the vast information needs and because of the large number of choice variables that seems inevitable. Because these choice variables are integer (it is not common to harvest part of a tree), extremely large combinatorial models with spatial and temporal interactions result. Consequently, we make many simplifying assumptions in this chapter, and an important part of our investigation is to explore the solvability of models with these types of choice variables.

Our approach tracks the location of all harvested and nonharvested trees so that location-specific solutions result and spatial relationships can be captured. Hann and Bare (1979:1) acknowledge these spatial relationships when they note that one property of uneven-aged stands is that they contain "trees of several ages that develop with significant interaction with surrounding trees of different ages." Although processes such as tree growth and mortality are spatially sensitive, the key spatial component we examine is regeneration. Because harvesting is scattered among the trees available for harvest, natural regeneration typically is relied on in harvest systems other than clearcutting. For this chapter, we assume that the area we are managing is generally conducive to natural regeneration and that it is completely relied on except under particularly poorly stocked initial conditions. Natural regeneration is spatially sensitive because proximity to seed trees often is necessary for successful regeneration after a tree's harvest. We assume that by the time a tree has reached a harvestable age, it has suppressed other competing trees in its immediate vicinity (including early regeneration). Typically this would be the result of crown dominance and the layer of duff that accumulates directly under and near a mature tree. Thus we assume that if a tree is harvested, a source for seeds (a seed tree) near the vacated area is needed for some period of time for regeneration to take place. In some cases, a reasonable regeneration lag may require actions (not modeled here) such as controlled burning to expose mineral soil for new seeds to germinate. We begin by presenting our integer programming formulation. Then we demonstrate the use of this formulation and explore its solvability with a simple case example.

Formulation

We assume that at commercial maturity, stem density is N trees per hectare, so we define the planning area as a grid of cells $1/N$ ha in size. We assume that at the minimum harvestable age, natural competition or precommercial thinning has created a density of 1 tree per cell and that this density remains constant throughout the remaining growth in the stand of trees. We realize that this simplistic characterization of tree density over time is not realistic, but it serves the purposes of this exploratory book. Accounting for a more complicated and realistic pattern of tree density over time is left as an extension (nontrivial) to this chapter.

With the spatial units (cells) so defined, we use the following formulation:

Maximize

$$\sum_t P_t Q_t - \sum_i \sum_j \sum_m C_m Y_{ijm}, \tag{4.1}$$

subject to

$$\sum_i \sum_j \sum_{m=1}^{M_{ij}} V_{ijmt} Y_{ijm} = Q_t \quad \forall t, \tag{4.2}$$

$$\sum_{hk \in \Omega_{ij}} \sum_{m=1}^{M_{hk}} S_{hkmt} Y_{hkm} \geq \sum_{m \in \theta_t} Y_{ijm} \quad \begin{matrix} \forall t \\ \forall i \\ \forall j, \end{matrix} \tag{4.3}$$

$$\sum_{m=1}^{M_{ij}} Y_{ijm} = 1 \quad \begin{matrix} \forall i \\ \forall j, \end{matrix} \tag{4.4}$$

$$Y_{ijm} \in \{0, 1\} \quad \forall i \qquad\qquad (4.5)$$
$$\forall j$$
$$\forall m,$$

with the following definitions:

Indexes

i indexes row, as does h,
j indexes column, as does k,
t indexes time period,
m indexes management prescription.

Variables

Y_{ijm} = the application (1) or nonapplication (0) of management prescription m to cell ij,
Q_t = the volume harvested in time period t.

Parameters

M_{ij} = the number of management prescriptions for cell ij,
M_{hk} = the number of management prescriptions for cell hk,
P_t = the price of timber volume in time t (discounted),
C_m = the cost of management prescription m (discounted),
V_{ijmt} = the timber volume obtained in time period t if management prescription m is implemented in cell ij,
S_{hkmt} = a parameter that is set at 1 if there is a seed-bearing tree in cell hk in time period t, if management prescription m is implemented, and 0 otherwise,
Ω_{ij} = the set of indexes for cells that are adjacent to and thus capable of seeding cell ij, within the time period after a harvest: Ω_{ij} = [$(i + 1, j)$, $(i - 1, j)$, $(i + 1, j + 1)$, $(i - 1, j + 1)$, $(i + 1, j - 1)$, $(i -1, j - 1)$, $(i, j + 1)$, $(i, j - 1)$],
θ_t = the set of indexes for management prescriptions that have harvests in time period t.

Equation (4.1) maximizes the discounted revenue of the volumes harvested less the harvesting costs. Equation (4.2) accumulates the volumes harvested in each time period into the Q_t variables. Equation (4.3) requires that if a cell (tree) is harvested in time period t, then an adjacent seed tree must be available to reseed it during that time period. Equation (4.4) forces one and only one management prescription to be selected for each cell. Equation (4.5) restricts the management choice variables (Y_{ijm}) to be binary.

We initially assume that the area is fully seeded at the beginning of the first time period so that the seeding constraint (4.3) is concerned only with the effects of harvesting. In the case example we show how we can relax this assumption and enable the model to track regeneration of initially unseeded cells as long as certain conditions are met. In many real-world situations, it may be necessary or desirable to artificially plant or seed a poorly stocked area before beginning a management regime that relies on natural regeneration. Even if natural regeneration is used, converting a poorly stocked area to one that can be modeled as we do here probably is best handled as a separate problem and is beyond the scope of this chapter. We can visualize using methods such as those in part II to approach such a conversion problem.

Case Example

We assume five time periods of 20 years each, with all management actions taking place at the beginning of each time period. This reflects a management situation in which we cut in a given stand only occasionally and must make long-term decisions on that basis. We also assume that trees begin to bear seeds (become seed trees) at age 30 and become available for commercial harvest at age 50. For each cell, we include nine prescription options (as in chapters 2 and 3) numbered in order: cut in time period 1, cut in time period 2, cut in time period 3, cut in time period 4, cut in time period 5, cut in time periods 1 and 4, cut in time periods 1 and 5, cut in time periods 2 and 5, and no harvest. For the net revenue objective function, we assume a discount rate of 4%, a stumpage price of $17.66/m^3, and a harvesting cost of $2 per tree.

We assume that a high probability of reseeding a harvested cell (tree area) is achieved in 20 years (one time period), with an average regeneration lag of 10 years, as long as an adjacent seed tree is available during that time period. For predicting yields, we assume that there is no harvestable volume before age 50, and after that the volume estimation amounts to a single-tree projection. We assume that at commercial maturity, stem density is 50 trees per ha, so we define the cell size to be 0.02 ha. For this case example, we adapted the yield function used in chapters 2 and 3 (from Rose and Chen 1977) based on our model assumptions to

$$V(t) = 4.206 \, (1 - e^{-.0301t})^{.6463}(1 - e^{-.0212t})^{1.243},$$

t = age in years,

$V(t)$ = volume per tree at age $t \geq 50$ years, in cubic meters,

for demonstrative purposes.

Results

Solvability

The first question that comes to mind with a formulation such as this is, "How large a model can be solved?" With integer choice variables defined at the scale of the space occupied by a single mature tree, our formulation is useful (even at the stand level) only if large models can be solved. Recently, heuristic solution techniques such as tabu search, genetic algorithms, and simulated annealing have been widely used to solve large (even very large) integer programs (see, for example, Murray and Church 1995). These techniques certainly would be applicable to our formulation and would be expected to provide good, feasible solutions for large models. However, with these techniques the analyst never knows for sure just how close to or far from optimality any feasible solution is unless an additional analysis such as some form of relaxation, empirical testing, or statistical inference (see Reeves 1993) is feasible and useful. Thus for this chapter we investigate the use of a branch-and-bound integer algorithm. Even when this approach does not arrive at a global optimum, it provides a bound on the suboptimization associated with any integer solution (based on linear programming solutions calculated during the branch-and-bound iterations).

To investigate the general solvability of our formulation, we built models with square land areas and initial ages determined by a uniform random number generator between 0 and 100 years. Note that this implicitly assumes that all cells are initially seeded (without regard to how the initial age distribution and seeding might have taken place). We were able to solve remarkably large integer models on this basis. The largest we were able to solve to a global optimum was a model with a 104×104 grid of cells, which implied 97,344 integer variables. The reason this problem is so solvable is that with this level of coefficient uniqueness, the optimal linear programming solution is integer or nearly so. In the following models, which have more redundancy in the initial conditions, obtaining optimal solutions was more difficult. Inclusion of non-declining yield constraints also made solution more difficult. Even in those cases, however, we arrived at solutions within 4% of the bound on optimality in a reasonable amount of computing time. Overall, this appears to be a very integer-friendly (Erkut et al. 1996) formulation, which would be helpful even if a given application was large enough to necessitate heuristic methods.

Solution Patterns

To investigate the solution patterns that result from using this formulation, we built models with a 12×12 grid of cells (1296 integer variables) and a variety

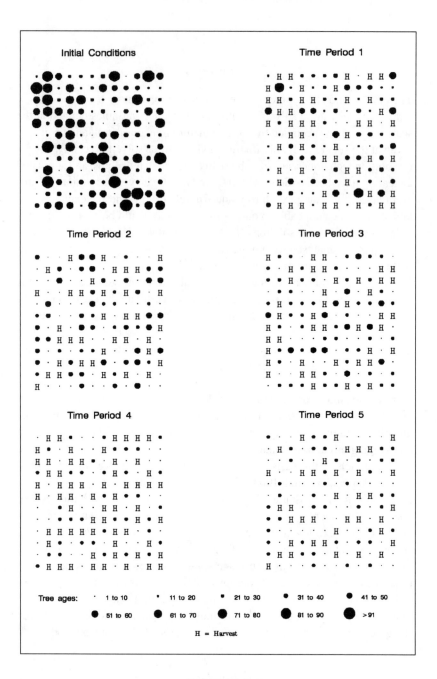

FIGURE 4.1

Model solution with uniform random initial ages.

of initial conditions (tree ages by cell). First, we built a model of this size with the uniform random age distribution between 0 and 100 years that we used to look at model solvability in the previous section. Figure 4.1 shows these initial conditions, along with the harvests and ages of nonharvested cells for time periods 1–5 from this solution. Table 4.1 presents the harvest volumes, objective function value, and bound (if applicable) for all the solutions discussed in this section. We should point out that the yield function used has a growth rate less than 4% at age 50, so harvesting is economically desirable at that age. With random initial conditions, harvests are placed sporadically over the stand throughout the planning horizon. The age structure evolves to a set of three ages (10, 30, and 50 years) as a byproduct of the time period length, and these ages are sporadically placed in the stand through 5 time periods. Even after the age structure is simplified, however, the volume (table 4.1) and number of trees (figure 4.1) harvested still fluctuate somewhat from period to period.

To further investigate this pattern, we applied a nondeclining yield constraint to this model:

$$Q_t \geq Q_{(t-1)} \quad \forall t \geq 2. \tag{4.6}$$

See table 4.1 for harvest and objective function results. We also solved a MAXMIN formulation (see Hof et al. 1986) that replaces our financial objective (4.1) with the following:

Maximize

$$\lambda \tag{4.7}$$

subject to

$$\lambda \leq Q_t \quad \forall t. \tag{4.8}$$

This maximizes the minimum harvest (λ) across the five time periods (again, see table 4.1 for results). In both of these solutions, nearly even flow (especially with the MAXMIN objective) is obtainable with a nearly even number of cells harvested in each of the latter time periods. And, there is a 9–12% loss in net revenue between the figure 4.1 solution and these solutions (table 4.1). Even more than our first solution in table 4.1, these two solutions approach Anderson and Bare's (1994:1758–1759) definition of "an uneven-aged stand in a steady-state condition" as one "with a diameter class structure such that at every cutting cycle a constant harvest can be removed in perpetuity while maintaining an invariant residual stand structure." Next, we explore several

TABLE 4.1
Solution Values for Figures 4.1–4.5

	Volume Harvested, by Time Period (m^3)					Net Revenue	Bound (if not optimal)
	1	2	3	4	5		
Figure 1 solution	144.01	94.08	105.54	139.91	80.81	$3816.41	—
Figure 1 solution with nondeclining yield constraints	108.15	109.98	112.31	112.54	121.39	$3347.91	$3492.71
Figure 1 solution with MAXMIN objective	113.83	113.86	114.18	114.94	114.27	$3481.06	114.32*
Figure 2 solution	248.75	36.54	42.84	233.99	47.08	$4996.46	$5206.15
Figure 3 solution	254.77	62.72	87.41	168.31	78.00	$5447.16	$5556.64
Figure 4 solution	65.35	5.62	68.91	56.92	91.35	$1532.49	$1553.58
Figure 5 solution	23.39	15.55	75.58	64.18	89.81	$ 948.00	—

*Applies to minimum harvest level, not net revenue.

systematic sets of initial conditions to test for departures from this approximate steady-state condition. In those tests, we do not apply nondeclining yield constraints or the MAXMIN objective but apply the original model formulation in equations (4.1)–(4.5).

Figure 4.2 presents a solution where the initial age of all the trees in the entire stand is 50 years. As one might expect, the solution cuts very heavily in the first time period. Then a cyclic pattern develops that is markedly different from the pattern in figure 4.1. In time periods 2 and 3, minimal cuts take place as the forest regrows after the initial harvests in period 1. Then, in time period 4, another large harvest takes place and the growth part of the cycle returns in period 5. This solution is thus a heavy harvest tempered only enough to leave a minimal number of seed trees for regeneration, followed by two periods of growth until a heavy harvest is again feasible. This might be called "nearly even-aged management" with shelterwood or seed tree cuts because it is even-aged except for the seed trees left.

We obtained a similar solution (not shown) with the initial ages set to 10 everywhere except all combinations of rows and columns 2, 5, 8, and 11. This represents the situation created if, 20 years ago, only a minimum number of 30-year old seed trees had been left. No cuts take place in time period 1 because there are no seed trees available to replace the extant ones. By time period 2, all the older trees are cut because the younger trees are now of seed tree age. Time period 3 cuts recreate a pattern much like the initial condition. By time period 5, a pattern much like that in time period 2 recurs. This cycle, which resembles the one seen in figure 4.2, shows that a similar pattern results whether initial stocking is near maximum or near minimum. These solutions could be constrained into a prespecified steady state, and an optimized conversion period would be calculated by the model, but it is noteworthy that the skewed numeric age distributions in these initial conditions tend to stay in a nearly even-aged and cyclic state. One can imagine a continuum of solutions that are anywhere between the steady state defined by Anderson and Bare (1994) and even-aged management, depending largely on the initial conditions.

In figure 4.3, we present a solution where the initial ages were set at 10 times the row number (numbered beginning at the top). This initial condition has a numeric age class distribution more like that in figure 4.1, but the spatial layout creates a very different optimal solution. A heavy cut is applied in time period 1 in the older cells (still leaving seed trees). Two rows (plus a few additional previous seed tree cells) are cut in time period 2 and again in time period 3. This cycle appears to start over in time period 4 with a heavy cut followed by two rows and some previous seed trees cut in time period 5. Despite its even initial numeric age distribution, this solution does not tend toward a steady state. Instead, it solves with a cycle of harvests whose spatial layout reflects the initial spatial age pattern.

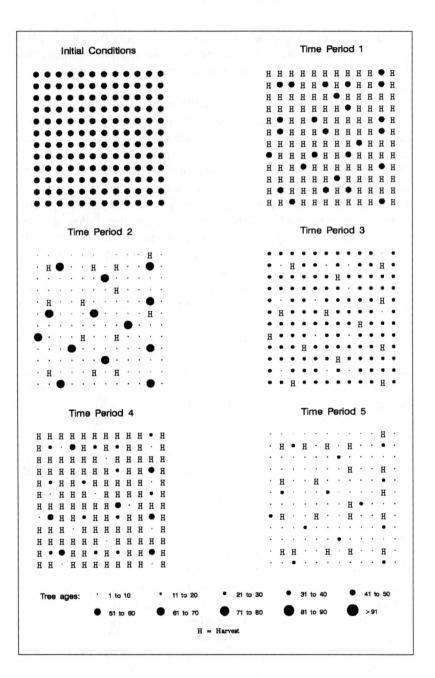

FIGURE 4.2

Model solution with all initial tree ages of 50 years.

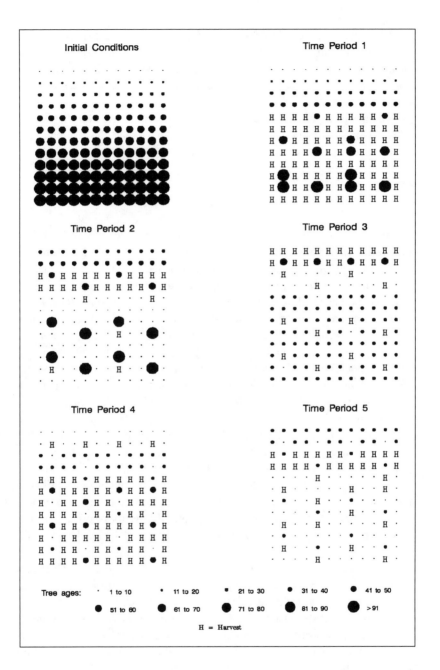

FIGURE 4.3

Model solution with initial tree ages set to 10 times the row number.

We should be cautious in drawing conclusions from these simple models with only five time periods, but most of our results do not display a simple steady state. Some managers may *desire* a steady state, but imposing it may not always be consistent with financial optimization or other forest management objectives. Most importantly, our results suggest that the spatial pattern of tree ages is important in addition to their numeric distribution. This point is further demonstrated in the next section, where we relax the assumption of a fully seeded initial stand.

Incomplete Initial Seeding

The previous solutions reflected initial conditions in which all cells had been previously seeded at the beginning of the planning horizon. This assumption can be relaxed with our formulation as long as the minimum harvest age is greater than the minimum seed tree age plus the regeneration period and as long as the initial conditions are consistent with the model assumptions. These assumptions are necessary to ensure that a tree needed to seed an unseeded cell will not be harvested before the regeneration period takes place and to avoid logical anomalies in the model. To see why these conditions are necessary, consider a cell with a 50-year-old tree and two empty cells to its right with no other seed trees that might regenerate them. This situation is not consistent with the model assumptions because the cell just to the right of the first cell should have been previously regenerated and should have an expected age of 10 years. As it is, the model could harvest the first cell if there is some other seed tree adjacent to it, so we cannot predict the age of the tree in the cell just to the right of the first cell. We can predict it if the initial conditions conform to the model assumptions. Under these conditions, the cell just to the right of the first one would have an initial age of 10, and the cell to its right would have an initial age of −30 (i.e., regeneration is still 30 years away). This is predictable because the minimum harvest age (50) is enough longer than the minimum seed tree age (30) to ensure the 20-year regeneration period (with a 10-year lag). The negative initial ages operate in our matrix generator, which sets ages in the integer program to zero until positive ages are reached.

To demonstrate how incomplete initial seeding affects model solutions, we solved the model depicted in figure 4.4, where the interior of the stand is seeded but the outer two layers are not. By setting ages to zero until regeneration occurs, we capture the radial expansion of tree regeneration along with the optimal harvesting strategy associated with it. In figures 4.4 and 4.5, the negative ages are shown for the initial conditions, and zero ages are shown in time periods 1 through 5 until growth begins, to portray the integer program solution. In time period 1, all of the interior is cut except to leave the necessary seed trees. Time period 2 has almost no harvest as the residual stand matures. In time period 3, a "ring" of cells is harvested. The interior is harvested again in time period 4, followed by a ring of cuts in time period 5. However, this ring is larger than the one in time period 3. Obviously, the center–ring–center cycle responds to the expansion of tree regeneration included in the model. Including enough time periods to see the long-term strategy is not feasible with our current solution software, but it seems likely that even after the entire area is stocked, this center–ring–center cycle may persist because of the age distribution created. This solution suggests that an expanding forest may be economically managed with a spatial cycle that expands with the forest. Of course, such a conclusion would be case-specific.

Our final solution is presented in figure 4.5. Here we demonstrate the use of the model, with negative ages to capture initially unseeded cells, for a case that is less systematic (and perhaps more realistic) than the previous ones. Here, the initial conditions reflect a poorly stocked stand with clusters of older trees in the southwest and southeast quarters, a few seed trees in the northwest quarter, and no stocking in the northeast quarter except a single 40-year-old seed tree. The negative initial ages are determined by the nearest seed source. Notice that the initial conditions are consistent with the model assumptions. In the first time period, the solution strategy is to harvest the older trees in the south, leaving seed trees as necessary. Then, in time period 2, these can be removed along with some of the trees in the northwest. In time period 3, two perimeter cuts are present in the south part of the area, followed by interior cuts in time period 4 and another set of perimeter cuts in time period 5. At the same time, a perimeter cut occurs in the northwest in time period 4, followed by an interior cut in time period 5. The basic pattern in figure 4.4 thus seems to occur in figure 4.5 at a smaller scale. From time period 3 on, the total volume harvested is fairly even (table 4.1), reflecting a cycle that cuts a relatively even number of trees in each of those time periods. In the trend shown in figure 4.5, the northeast quarter is regenerated gradually. The solution seems reasonable but would not be intuitively obvious without the spatial optimization analysis. As before, the initial conditions spatially affect the solution through all five time periods.

Perhaps the most surprising result we report from the case example exercise is the solvability of this model formulation. The prospect of solving a spatially explicit problem with 14,000 integer variables could cause an analyst to give up before starting—at least when using a branch-and-bound algorithm, seeking a global optimum. Beyond that, the case example shows that solutions to this type of problem may suggest management strategies that would never have come to mind without such an analysis. Depending on the initial

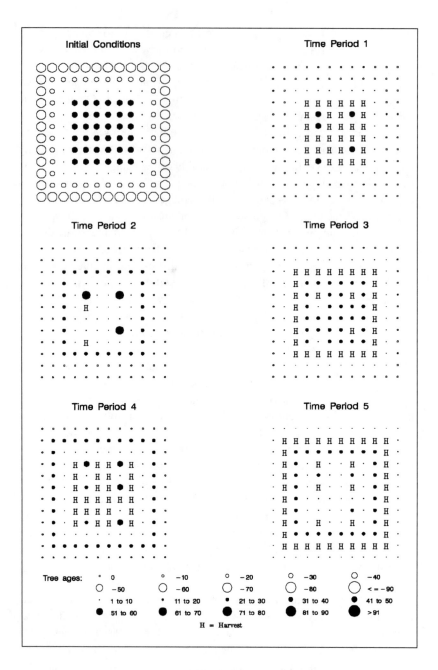

FIGURE 4.4

Model solution with initial conditions of 50-year-old trees in interior,
surrounded by poor stocking.

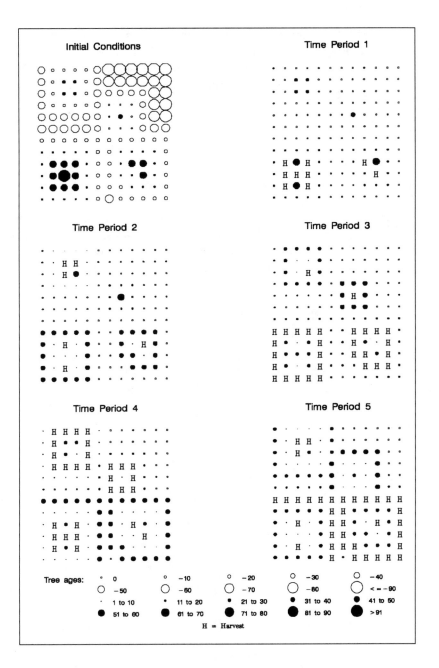

FIGURE 4.5

Model solution with nonsystematic initial conditions reflecting
poor stocking through much of the planning area.

conditions, optimal solutions to this problem may or may not tend toward a simple steady state. Our solutions suggest that the location of tree ages is important, in addition to their numeric distribution. Also, a fixed cutting cycle length and even flow may not be financially optimal, although they may be desired for other reasons. If spatial considerations for wildlife, water, aesthetic, or other resources are important, they must be directly accounted for in the model, as discussed elsewhere in this book. Our models in part II might also suggest an extension of this model to include seed trees farther away from harvested cells, with diminishing probabilities of success as a function of distance. Conceptually, our model is not scale dependent and might prove useful for larger multistand problems, subject to model solvability. The data requirements for a model such as this are substantial, however, which might limit the scale of application before model solvability does. In chapter 5 (the last in part I), we develop simple proximity relationships from a simulation model and apply them in a real-world case study.

5

COMBINING SIMULATION
WITH OPTIMIZATION: HABITAT PLACEMENT
FOR THE NORTHERN SPOTTED OWL

Simulation methods have been widely applied in ecology, whereas optimization methods have been more common in economics and management science. Perhaps because of these disciplinary loyalties, simulation and optimization methods sometimes are regarded competitively, and each discipline may not have given enough attention to the methods typically used by other disciplines. Most would agree that in modeling ecological systems, simulation models can include much more detailed (and realistic) biological behavior than optimization models can and still be solvable (Dunning et al. 1995; Turner et al. 1995). At the same time, few could deny that optimization models are intrinsically more powerful in finding efficient resource allocations and performing tradeoff analyses between competing outputs or system components such as timber production and wildlife habitat.

Simulation models allow a land manager to explore the implications of alternative large-scale land management scenarios for which landscape-level experiments are not feasible. Questions about the effects of both the amount and the arrangement of habitat patches are of special interest here and are amenable to analysis with spatially explicit population models (Dunning et al. 1995). These simulation models help the analyst consider how fragmentation, size and shape of patches, and connectivity or isolation of patches affect animal populations and are useful tools for evaluating and testing alternative

This chapter was adapted from J. Hof and M. G. Raphael, Optimization of habitat placement: A case study of the northern spotted owl in the Olympic Peninsula, *Ecological Applications* 7, no. 4 (1997): 1160–1169, with permission from the publisher, the Ecological Society of America.

landscape designs. Optimization models, on the other hand, can be design tools that aid in formulating these alternative scenarios. Whereas a simulation model can be used to evaluate an array of alternative plans, an optimization model can be used proactively to design a plan. The purpose of this chapter is to develop an optimization procedure that can be constructed from proximity relationships and used in concert with simulation methods, capturing the strengths of both in a spatial analysis. We focus on the northern spotted owl (*Strix occidentalis caurina*) in the Olympic Peninsula as a case study. We use a simulation model to estimate relationships for constructing an optimization model with a simplified formulation. We then use the simulation model to check the conclusions drawn from optimization, thereby recapturing the biological detail and reliability that might be compromised in the simplified optimization formulation.

The Northern Spotted Owl

For more than a decade, conservation of the northern spotted owl, hereafter referred to simply as "the owl," has been a central issue in forest management in the Pacific Northwest (Thomas et al. 1990, 1993; U.S. Department of Agriculture and U.S. Department of the Interior 1994; Forsman et al. 1996). The owl was listed as a threatened species in 1990 (U.S. Department of the Interior 1990), and a strategy for managing late-successional and old-growth forests on federal lands in the range of the owl was adopted in 1994 (U.S. Department of Agriculture and U.S. Department of the Interior 1994). Although the underlying strategy for owl conservation was developed using criteria derived from simulation modeling (Lamberson et al. 1992, 1994), the Northwest Forest Plan (NFP) originally was designed largely without the use of any formal modeling effort. After completion of the NFP (FEMAT 1993) as part of the analyses in support of a final Environmental Impact Statement, Raphael et al. (1994) used a spatially explicit simulation model ("OWL," McKelvey et al. 1992) to compare how owl populations might respond to land allocations under the NFP and under two alternative plans. In a subsequent analysis, this simulation model was used to compare alternative scenarios for management of owl habitat on the nonfederal land on the Olympic Peninsula (Holthausen et al. 1995). In 1995, an alternative was considered by the U.S. Fish and Wildlife Service that included retaining a specially designated area of nonfederal habitat on the western border of the federal lands (U.S. Department of the Interior 1995). This alternative, hereafter referred to as "the plan," is analyzed in this chapter. Following Holthausen et al. (1995), we divide the Olympic Peninsula into hexagonal 1500-ha cells, the size of which represents a core area or territory for an owl pair. This yields a model with 1681 cells

(41 × 41), with the carrying capacity set at zero for any cell that does not currently contain any habitat.

To simulate the spotted owl population, we (Hof and Raphael 1997) used OWL (version 2.01, described by McKelvey et al. 1992). The explicit spatial content of this model is important because the owl's habitat needs are specific to late-successional and old-growth forests, and fragmentation or connectedness of that habitat may be important for owl survival and population size. Our study focused on the Olympic Peninsula in Washington. This peninsula has been identified as an area of special concern because the owls on the peninsula probably have very little interaction with owls in the remainder of the range (Thomas et al. 1990; U.S. Department of Agriculture 1988, 1992; U.S. Department of the Interior 1992).

The Model

The basic variables in this problem are the amount of habitat in each cell that is retained (or protected). Because habitat for the northern spotted owl is low-elevation late-successional and old-growth forests, which take hundreds of years to develop, the decision to retain habitat in any cell or not is semipermanent. This actually simplifies the problem relative to a situation in which habitat status changes more rapidly and habitat development scheduling over time is necessary. Thus we use a static optimization model to take advantage of this simplifying characteristic. This implies that the dynamic processes of birth, dispersal, and death modeled in the simulation model will be captured in an equilibrium sense with static relationships in the optimization model.

The dynamics of owl population growth and dispersal captured in the simulation model are complex, but we assume that at any point in time, the population in any cell is limited by either the amount of habitat, the connectivity to other cells' owl populations, or both. A simple characterization of connectivity would be that each cell's population is potentially limited by the population in the cells immediately surrounding it. For each cell i, we define the set of immediately surrounding cells as set Ω_i. Our simple proximity relationship is analogous to a first-order spatial autocorrelation, and because each cell in each Ω_i is affected by its surrounding cells, this relationship implies a cascading effect beyond the Ω_i set for each cell i. The combination of potentially limiting factors could then be formulated as

$$p_i \leq f\left(\sum_{k \in \Omega_i} p_k\right) \quad \forall i$$

$$p_i \leq g\left(C_i\right) \quad \forall i,$$

where

i indexes the I cells,

p_i = the population in the ith cell,

Ω_i = the set of cell indexes that surround cell i,

C_i = the proportion of cell i that is retained as habitat,

$\sum_{k \in \Omega_i} p_k$ = the total population in all of the cells in set Ω_i,

f and g = empirical functions.

The f and g functions do not describe dynamic processes but are static, deterministic relationships estimated from empirical or simulated observations. In general the f and g functions would be nonlinear but could be approximated in a linear optimization model as a series of linear line segments as long as they universally exhibit diminishing returns. This piecewise approximation is accomplished as follows, using a g function as an example (see figure 5.1). We know that the range of C_i is from 0 to 1. We then divide C_i into H segments, each Q_h in size. If we have five equal-sized segments, then all the Q_h are 0.2. If we define a variable for each segment of C_i as Y_{ih}, then

$$Y_{ih} \leq Q_h \quad \forall h,$$

and C_i is determined by

$$C_i = \sum_{h=1}^{H} Y_{ih}.$$

The g function is then approximated by

$$p_i \leq \sum_{h=1}^{H} b_h Y_{ih},$$

where b_h approximates the g function for segment h of C_i and is defined here as the average slope of the g function between the left and right end points of segment h.

Thus a model to maximize owl population is as follows:

Maximize

$$\sum_{i=1}^{I} p_i, \tag{5.1}$$

$$p_i \leq \sum_{j=1}^{J} a_j X_{ij} \quad \forall i, \tag{5.2}$$

$$X_{ij} \leq S_j \quad \forall i \tag{5.3}$$
$$\forall j,$$

$$\sum_{j=1}^{J} X_{ij} = \sum_{k \in \Omega_i} p_k \quad \forall i, \tag{5.4}$$

$$p_i \leq \sum_{h=1}^{H} b_h Y_{ih} \quad \forall i, \tag{5.5}$$

$$Y_{ih} \leq Q_h \quad \forall i \tag{5.6}$$
$$\forall h,$$

$$\sum_{h=1}^{H} Y_{ih} = C_i \quad \forall i, \tag{5.7}$$

$$0 \leq C_i \leq B_i \quad \forall i, \tag{5.8}$$

where

j indexes the J segments that approximate f,

h indexes the H segments that approximate g,

a_j = a coefficient that approximates the f function for the jth segment of $\sum_{k \in \Omega_i} p$

b_h = a coefficient that approximates the g function for the hth segment of C_i,

X_{ij} = the jth segment of $\sum_{k \in \Omega_i} p_k$,

Y_{ih} = the hth segment of C_i,

S_j = the maximum segment size of X_{ij},

Q_h = the maximum segment size of Y_{ih},

B_i = the proportion of habitat currently existing in cell i, and thus available for retention.

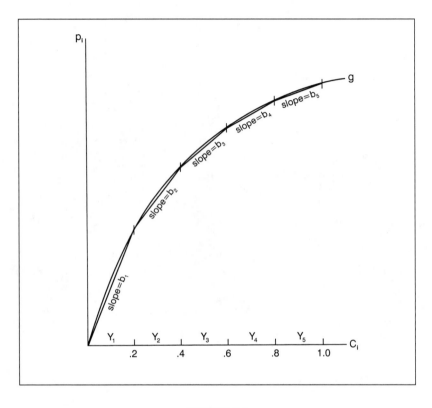

FIGURE 5.1
An example of the piecewise linear approximation method
used in the optimization model.

Equation (5.1) is simply an owl population–maximizing objective function. In some parts of the analysis that follows, we will use the following objective function:

Maximize

$$1000 \left(\sum_{i=1}^{I} p_i \right) + \sum_{i=1}^{I} (1 - C_i), \tag{5.9}$$

which maximizes the owl population (weighted with an arbitrarily large number set so that owl population is not compromised), with no more habitat area than is necessary. Equation (5.2) approximates the connectivity functions (f) with the X_{ij} segments, defined in equations (5.3) and (5.4). Equation (5.5) approximates the carrying capacity functions (g) with the Y_{ih} segments, defined in equations (5.6) and (5.7). Equation (5.8) limits the habitat retention variables, C_i, to be less than or equal to the amount of habitat in each cell that is available for retention (B_i). The expected population in each cell (p_i) is determined by whichever (possibly both) of equations (5.2) and (5.5) is limiting. For the case study in this chapter, with 1681 cells, this formulation yields a model with approximately 20,000 variables and 12,000 equations.

To bracket the likely actual population behavior of the owl, we included two rule sets from Holthausen et al. (1995): rule set b, which is somewhat pessimistic, and rule set d, which is somewhat optimistic (relative to projected rates of population change). These rule sets contain parameter estimates for adult and juvenile survival rates and fecundity in relation to the amount of habitat in each cell. We then derived a connectivity (f) function and a carrying capacity (g) function for each of those rule sets, based on the simulation model data and results.

The Connectivity (f) Function

To estimate the connectivity (f) function, we began by selecting cells from the study area where spotted owls were least likely to be habitat limited. As discussed in the next section, available evidence suggests that owl survival (and hence likelihood of occupancy of a cell) does not increase when habitat exceeds 60% of a cell (Bart and Forsman 1992; Bart 1995). There were 100 cells that had > 60% habitat, which we included for the function estimation. Given that these 100 cells are not habitat limited, we assume that the populations observed in the simulation runs are limited (determined) by the presence of population in surrounding cells. Next, for each rule set (b and d) we ran 50 replications of 100-year simulations. We estimated the population of each cell as mean occupancy, the number of years the cell was occupied by a pair of owls divided by the total years in the simulation (100). Using average simulated population numbers across all replications for each of the 100 selected cells, we then regressed each cell's mean population against the total mean population of the six cells surrounding it. We tried several functional forms but settled on a log–log transformation that yielded an exponential function.

The regressions obtained in this fashion for the two rule sets are as follows:

Rule Set b

$$p_i = 0.3464 \left(\sum_{k \in \Omega_i} p_k \right)^{0.507},$$

F value $= 872.21$ ($\alpha \leq 0.0001$),

Adj. $R^2 = 0.898$.

Rule Set d

$$p_i = 0.4272 \left(\sum_{k \in \Omega_i} p_k \right)^{0.445},$$

F value $= 902.06 \qquad (\alpha \leq 0.0001),$

Adj. $R^2 = 0.901.$

Obviously, the statistical confidence in these equations is quite good. We also inspected plots to ensure that the functions were reasonable relative to the plotted data.

These functions were then approximated in equations (5.2)–(5.4) with the following parameters:

		Rule Set b	Rule Set d
a_1	=	0.766	1.044
a_2	=	0.272	0.312
a_3	=	0.164	0.176
a_4	=	0.112	0.124
a_5	=	0.090	0.097
S_1	=	0.2	0.2
S_2	=	0.5	0.5
S_3	=	1.0	1.0
S_4	=	1.0	1.0
S_5	=	∞	∞

The Carrying Capacity (g) Function

As with the connectivity (f) functions, we used simulation model results to estimate the carrying capacity (g) functions. We plotted mean owl occupancy per cell against percentage habitat in each cell from simulation replications based on retention of all current habitat, thus maximizing connectivity within the limits of the study area. For rule b, the plots appeared nearly linear from the origin to a point where habitat was 80% and mean occupancy was .75 pairs. Mean occupancy appeared constant for habitat percentages beyond 80%. We therefore used only two segments to approximate this relationship. For rule d, the plot appeared to have more curvature (diminishing marginal occupancy as a function of percentage habitat). Thus we approximated this relationship with three segments that all have diminishing but nonzero slopes. On this basis, the parameters in equations (5.5)–(5.7) were set as follows:

Rule Set b		Rule Set d
b_1 =	0.875	1.75
b_2 =	0	1.125
b_3 =	N/A	0.125
Q_1 =	0.8	0.2
Q_2 =	0.2	0.4
Q_3 =	N/A	0.4

We did not statistically regress these relationships because of the logical requirements that the intercept term be nonnegative and the function exhibit diminishing returns. Based on visual examination of the plots, the functions assumed are good approximations, but it is possible that the carrying capacity (g) functions are somewhat sigmoidal (e.g., logistic) with small marginal capacities associated with lower percentages of habitat, medium marginal capacities in the midrange of habitat percentages, and then diminishing marginal capacities at the higher habitat percentages. Unfortunately, this functional form would create a nonconvex program in the optimization model, and the approximation of the carrying capacity function would require integer methods (mixed-integer programming or separable programming). If one tried to solve such a nonconvex program with linear methods, the approximation segments (the Ys in figure 5.1) would not enter solution in the correct order (the b coefficients in figure 5.1 would be improperly ordered). This is a good example of the simplifications necessary to make the optimization model solvable, even given the basic simplicity of the formulation.

Results

Comparison of the Simulation Model with the Optimization Model

We initially compare the simulation model with the optimization model, with all current habitat available. Plate 1 (top) shows the owl habitat currently extant and thus available for retention (the B_i coefficients in equation (5.8)). This includes 328,970 ha of habitat, distributed across the peninsula but concentrated in the federal ownerships. Plate 1 (bottom) also shows the habitat proposed to be retained in the plan. Plate 2 shows the simulation model projections of the owl population if all of the available habitat is retained, with rule sets b (top) and d (bottom). The simulation model predicts a mean population of 177 pairs under rule set b and 285 pairs under rule set d. When we solved the optimization model with constraints (5.2)–(5.8) and objective function (5.9) and with all current habitat available, we obtained a population of

177 pairs under rule set b and 287 pairs under rule set d. The optimized mean population layouts for rule sets b (top) and d (bottom) are shown in plate 3. With rule set b, 303,000 ha of the 328,970 ha of habitat is used; with rule set d, 304,020 ha of habitat is used. As noted previously, objective function (5.9) maximizes the owl population but with a minimum of habitat.

The close similarity in the magnitude of the population projections between the simulation and optimization models was expected because we used the simulation model to construct the carrying capacity functions in the optimization model. The spatial distributions of population, which are determined in the optimization model primarily by the connectivity functions (given a particular habitat layout), are not quite as similar. The connectivity functions are limiting in the optimization model, as evidenced by the fact that not all of the 328,970 ha of available habitat was used in maximizing owl population with either rule set b or rule set d. The population distributions from the simulation model (plate 2) are more concentrated and contiguous near the center of the peninsula than the population distributions from the optimization model (plate 3). This results from a basic difference in how connectivity is handled in the two models. In the simulation model, owl behavior includes a nonrandom ability to locate habitat within certain distances (specified by the rule set). In contrast, the static optimization model formulation accounts for connectivity as a simpler proximity relationship. This formulation, as an analogue to a first-order spatial autocorrelation, implicitly treats owl dispersal as a random, directionless process that entails no selective behavior on the part of the owl. Thus the optimization model is more conservative in that a high degree of connectivity for any given cell requires surrounding habitat in all directions. This logical difference between the two models persists, even though the simulation model was used to determine the particular parameter values for the optimization model.

One other difference between the two models is noteworthy. In the optimization model formulation, if no habitat is present in a cell, population is forced to zero by equations (5.5)–(5.7). In the simulation model, however, some small mean population may be indicated in nonhabitat cells because across the simulation time periods and replications, pairs of owls may encounter and reside in nonhabitat cells for some time.

In the next section, we perform an experiment on the habitat layout of the owl plan to see if, despite these differences, the models can be used together.

Evaluation of the Plan

To begin our optimization analysis of the plan, we imposed its habitat layout (plate 1, bottom) on the optimization model to observe the predicted owl population. We accomplished this by converting equation (5.7) into an equality and by replacing the C_i in equation (5.7) with the plan habitat levels per cell. Also, because the habitat layout is fixed in these solutions, we used objective function (5.1). The results are shown in plate 4 for rule sets b (top) and d (bottom). The plan habitat layout (plate 1, bottom) includes 285,750 ha of habitat, concentrated in the federal ownerships (which happen to occupy the center of the peninsula) and in nonfederal lands on the western border of the federal land. The resulting population projections in plate 4 show a population of 152 pairs for rule set b and 242 pairs for rule set d, which is a loss relative to those in plate 3 (as expected from a 43,220-ha reduction in habitat). By far, the majority of this population loss occurs around the perimeter of the peninsula, where the plan does not retain habitat. Otherwise, the spatial population distributions in plates 3 and 4 are similar for each rule set.

When the plan habitat layout is imposed on the simulation model, the resulting mean population (again, across 50 replications and 100 years in each replication) is 152 pairs with rule set b and 250 pairs with rule set d. The spatial distributions of these mean population simulations are given in plate 5. These projections also suggest a loss of population magnitude relative to plate 2 (again, as expected), but the projected population distributions across the peninsula from imposing the plan on the simulation model are quite similar to those in plate 2 for each rule set. Apparently, because of the ability of the owl to find habitat modeled in the simulation model, the loss of perimeter habitat in the plan does not significantly affect the spatial distribution of population across the peninsula (even around the perimeter).

Next, we tested whether the optimization model could be used to select habitat for retention and whether this optimized solution would produce a larger owl population for the given amount of habitat than the proposed habitat management plan. For this analysis we solved the optimization model with the original equation (5.7) and the equation (5.1) objective function but also with an additional constraint:

$$\sum_{i=1}^{I} C_i \leq 285{,}750. \tag{5.10}$$

This limits the total amount of habitat allowed to the amount in the plan but allows the optimization model to optimally place that limited amount of habitat. The optimized habitat layouts (plate 6) are very different from the plan; the optimized habitat layouts are relatively decentralized, though still very well connected, for both rule sets. This, again, reflects the highly random definition of dispersal and connectivity implicit in the static optimization formulation. The projected population (plate 7) with rule set b is back up to 167 pairs, and with rule set d the population is back up to 281 pairs. Thus if the optimization model formulation is realistic, significant gains in expected owl population might be gained if we could exchange some perimeter habitat for some of the central habitat in the plan.

Because this conclusion clearly depends on the optimization model formulation, we next imposed the optimized habitat layouts in plate 6 on the simulation model (with the appropriate rule sets applied). The resulting simulated mean population distributions are shown in plate 8 for rules sets b (top) and d (bottom). The resulting mean populations (across 50 replications and 100 years per replication) were 153 owl pairs with rule set b and 263 owl pairs with rule set d. Comparing these numbers with the mean populations from the simulations with the plan habitat layout indicates an insignificant increase of one pair from the optimization for rule set b and a more significant increase of 13 pairs for rule set d. Thus, especially with rule set d, the optimized habitat layout may be superior to the plan layout, even according to the simulation model. Again the spatial distributions of mean population from these simulations across the peninsula (plate 8) are quite similar to the maps in plate 2 (and in plate 5). The simulation model results are not very spatially sensitive to the different habitat layouts in plates 1 and 6 because all these include the central habitat area and because of the ability of the simulated owls to locate habitat. The optimization model accounts for no tendency on the owls' part to disperse toward suitable habitat and is thus more spatially sensitive, creating a greater population difference between the plan layout and the optimized layout.

Our analysis suggests that the owl population might be enhanced if additional perimeter (nonfederal) habitat were traded for central (federal) habitat. Realistically, however, the possibility of trading federal habitat for additional nonfederal habitat probably is remote. And according to the results from the simulation model, the gains in terms of expected owl population would be modest under rule set d and insignificant under rule set b. The desirability of some habitat decentralization for a capable disperser, as our optimization results suggest for the more optimistic rule set d, has been noted elsewhere in the literature (e.g., With and Crist 1995). In chapter 13 we discuss a theoretical hypothesis that as populations approach extinction thresholds (consistent with the more pessimistic rule set b), habitat concentration becomes more and more desirable.

We should point out that the spatial pattern of ownership in the Olympic Peninsula is unique. In cases where federal ownership does not happen to occupy the center of the planning area, ownership patterns may be less useful in defining a wildlife habitat layout plan (and the optimization analysis might be more useful). Our optimization model results might be given more weight if, as we learn more about the behavior of the owl, we discover that its dispersal and connectivity are more random than the Holthausen et al. (1995) simulation model reflects. This optimization formulation might also be more advantageous for animals that exhibit highly random dispersal behavior or are less capable of dispersal than the northern spotted owl. In part II, reproduction and dispersal dynamics are modeled explicitly.

PART II

REACTION–DIFFUSION MODELS

One of the most fundamental ecological processes that occurs at the landscape scale is the spatial dispersal of organisms over time. In this part, we focus on the strategic placement (usually through protection) of habitat, taking the dispersal behavior of plants and animals into account. This problem is conducive to optimization because areas that are capable of providing habitat typically are a scarce resource, and the habitat placement decision typically determines not only the amount of habitat protected but also its fragmentation.

Fragmented landscapes often occur as habitable areas of varying size and shape, embedded with varying degrees of compaction in a matrix that is largely inhospitable. Consequently, the fragmentation of a particular habitat complex and the manner in which organisms disperse strongly interact to affect population abundance and distribution. The importance of population dispersal has led to the development of spatially explicit models of population dynamics in recent decades. Among the earliest works in this field are the pioneering studies on random dispersal by J. G. Skellam (1951), Kierstead and Slobodkin (1953), and Segel and Jackson (1972). Their reaction–diffusion methods are based on the probabilistic concept of random walks (Fisher 1937) or quasirandom walks undertaken by dispersing organisms. In the aggregate, these dispersal assumptions lead to deterministic population diffusion models similar to molecular diffusion models from physics (see Pielou 1977; Okubo 1980).

Most of this introduction was adapted from M. Bevers and C. H. Flather, Numerically exploring habitat fragmentation effects on populations using cell-based coupled map lattices, *Theoretical Population Biology* 55 (1999): 61–76, with permission from the publisher, Academic Press.

Populations in realistically fragmented landscapes have proved to be difficult to study with reaction–diffusion methods, which typically involve continuous (over time and space) partial differential equations to describe rates of population change (Holmes et al. 1994). Although continuous variable calculus has analytical advantages in general, it imposes important modeling limitations for these fragmented habitats (Levin 1974). The extension by Levin (1974), Allen (1987), and others of single-patch reaction–diffusion models to multipatch formulations converts the model from a single partial differential equation to a system of ordinary differential equations. This multipatch approach allows the examination of habitat complexes and is the spatial basis for our models in chapters 8 and 9. In chapters 6, 7, and 10, we subdivide patches into smaller units (uniform cells; see Kaneko 1993), which allows us to account for variations in population density within each patch as well as retaining the patch-level information.

For species that establish and defend distinct breeding territories, those territories define the spatial scale at which reproduction and dispersal processes occur within the population. Discretizing habitat with a grid of appropriately sized cells converts the habitat complex into a lattice representation based on those breeding territories. For more continuously distributed species, any habitat approximation error resulting from the use of cells can be reduced by refining grid resolution, as in finite numerical approximation methods.

By further discretizing reaction–diffusion models with respect to time, we convert the system of ordinary differential equations into an enlarged system of difference equations that may offer a more realistic formulation for many species (Levin 1974; Holmes et al. 1994). In particular, discrete time intervals may be more appropriate than continuous time modeling for species with distinct breeding and dispersal periods. For other species, discrete time steps still offer many computational advantages (Cushing 1988). Next, we derive our basic discrete reaction–diffusion model from the classic model.

Allen (1987, following Levin 1974) describes a random walk (passive) reaction–diffusion model for a fragmented complex of N habitat areas (we apply it to both patches and cells) as

$$\frac{dv_i}{dt} = v_i f_i(v_i) + \sum_{j=0}^{N} D_{ij}(v_j - v_i) \qquad i = 1, \ldots, N, \tag{II.1}$$

$$D_{ij} = D_{ji} \geq 0 \quad \forall i, j,$$

$$v_0 = 0,$$

where v_i is the population in area i as a function of time t, $f_i(v_i)$ is the per capita rate of reproduction (net of mortality unassociated with dispersal, as discussed later), and D_{ij} is defined as an area-to-area passive diffusivity constant that determines the net area i population gain from (or loss to) area j based on the difference between the area populations. Nonhabitat is indexed by $j = 0$. In this model, dispersing organisms can successfully cross regions of nonhabitat, but some of them perish and are treated as dispersers into nonhabitat. Because the nonhabitat population v_0 is fixed at zero, $D_{i0} \times -v_i$ defines the population loss caused by unsuccessful dispersal from any area i. The areas (indexed by i) that are "connected" by positive diffusivity ($D_{ij} > 0$) with at least one other area (some $j \neq i > 0$) form the habitat complex. This definition of habitat area connectivity imposes no constraint on the particular arrangement of habitat in a complex.

Expanding the diffusion summation in equation (II.1) and recollecting the terms into separate summations for emigration and immigration, we get

$$\frac{dv_i}{dt} = v_i f_i(v_i) - \sum_{\substack{j=0 \\ j \neq i}}^{N} D_{ij} v_i + \sum_{\substack{j=1 \\ j \neq i}}^{N} D_{ji} v_j \quad \forall i, \tag{II.2}$$

where we now distinguish the diffusivity constants D_{ij} and D_{ji} to indicate the direction of dispersal.

Reproduction and diffusion are modeled as simultaneous processes in this equation, so that reproduction in each area can implicitly contribute to the diffusion summation terms. In a simple age-structured population comprising only adults and juveniles, four cases are possible: only adults disperse, only juveniles disperse, adults and juveniles disperse identically, or adults and juveniles disperse but not identically. To clarify this, we define g_{ij} as the proportion of potential dispersers in area i expected to move to area j per unit of time, so that g_{i0} is the proportion that perish in nonhabitat while dispersing from area i per unit of time, and $g_{ii} = [1 - \sum_{j=0, j \neq i}^{N} g_{ij}]$ is the proportion expected to remain in area i per unit of time. Then, if only adults disperse, equation (II.2) becomes

$$\frac{dv_i}{dt} = v_i f_i(v_i) - \sum_{\substack{j=0 \\ j \neq i}}^{N} v_i g_{ij} + \sum_{\substack{j=1 \\ j \neq i}}^{N} v_j g_{ji} \quad \forall i,$$

and we see that all $D_{ij} = g_{ij}$. If only juveniles disperse, equation (II.2) becomes

$$\frac{dv_i}{dt} = v_i f_i(v_i) - \sum_{\substack{j=0 \\ j \neq i}}^{N} v_i f_i(v_i) g_{ij} + \sum_{\substack{j=1 \\ j \neq i}}^{N} v_j f_j(v_j) g_{ji} \quad \forall i,$$

and we see that all $D_{ij} = f_i(v_i)g_{ij}$. This is equivalent to

$$\frac{dv_i}{dt} = v_i f_i(v_i)\left(1 - \sum_{\substack{j=0 \\ j\neq i}}^{N} g_{ij}\right) + \sum_{\substack{j=1 \\ j\neq i}}^{N} v_j f_j(v_j)g_{ji} \quad \forall i,$$

which simplifies to

$$\frac{dv_i}{dt} = \sum_{j=1}^{N} v_j f_j(v_j)g_{ji} \quad \forall i.$$

If both adults and juveniles disperse identically, equation (II.2) becomes

$$\frac{dv_i}{dt} = \sum_{j=1}^{N} v_j f_j(v_j)g_{ji} - \sum_{\substack{j=0 \\ j\neq i}}^{N} v_i g_{ij} + \sum_{\substack{j=1 \\ j\neq i}}^{N} v_j g_{ji} \quad \forall i,$$

or

$$\frac{dv_i}{dt} = v_i f_i(v_i) - \sum_{\substack{j=0 \\ j\neq i}}^{N} v_i \left[1 + f_i(v_i)\right] g_{ij} \tag{II.3}$$

$$+ \sum_{\substack{j=1 \\ j\neq i}}^{N} v_j \left[1 + f_j(v_j)\right] g_{ji} \quad \forall i,$$

and we see that all $D_{ij} = [1 + f_i(v_i)]g_{ij}$. We note that a restriction $g_{ij} = g_{ji}$ is not needed because we have decoupled immigration from emigration. Having eliminated the reliance on population gradients, we can model less restrictive types of diffusion (as in Holmes et al. 1994). For the latter two cases in which juveniles are dispersing with stationary probabilities, we also note that the D_{ij} diffusivity parameters in gradient-based models should be treated as constants only when the net per capita area reproduction rate functions (f) are also constant.

Converting the single-stage population system of equation (II.3) to discrete time results in

$$v_{it} = v_{i(t-1)} + v_{i(t-1)} f_i(v_{i(t-1)}) - \sum_{\substack{j=0 \\ j\neq i}}^{N} v_{i(t-1)} \left[1 + f_i(v_{i(t-1)})\right] g_{ij}$$

$$+ \sum_{\substack{j=1 \\ j\neq i}}^{N} v_{j(t-1)} \left[1 + f_j(v_{j(t-1)})\right] g_{ji} \quad \forall i, t$$

plus a finite difference approximation error term. For species with distinct breeding seasons, an approximation error would be more correctly assigned to equation (II.3). The discrete time equation simplifies to

$$v_{it} = \sum_{j=1}^{N} \left[1 + f_j(v_{j(t-1)}) \right] v_{j(t-1)} g_{ji} \quad \forall i, t. \tag{II.4}$$

Equation (II.4) has also been used by Kot and Schaffer (1986) as the foundation for integrating to a discrete time, continuous space formulation. In discrete time simulation analysis, the values f_j (net reproduction) and g_{ji} (dispersal) could be defined as random functions $f_j(V, H, t-1)$ and $g_{ji}(V, H, t-1)$, respectively, where V is the vector of area populations and H is a matrix of habitat area conditions in the complex. Such functions could be systematically as well as stochastically time variant.

For species that mature in a single breeding season and whose adults and juveniles both disperse but not identically, we can derive the following equation in place of equation (II.4):

$$v_{it} = \sum_{j=1}^{N} v_{j(t-1)} f_j^A(v_{j(t-1)}) \left[g_{ji}^A + f_j^J(v_{j(t-1)}) g_{ji}^J \right] \quad \forall i, t.$$

Here we have partitioned net reproduction f_j and dispersal g_{ji} into separate terms for adult survivorship f_j^A and dispersal g_{ji}^A and for net natality f_j^J and juvenile dispersal g_{ji}^J. Multiple life stages could also be used to depict more detailed reproduction and dispersal process (van den Bosch et al. 1992; see chapter 9).

In this part we typically combine our reaction–diffusion formulation with limits on carrying capacity, where the limit is affected by the decision to protect or otherwise manage habitat. Following reproduction and dispersal throughout the complex in each time step, we treat area populations from equation (II.4) in excess of carrying capacity as mortality resulting from dispersal into fully occupied territories. This can also be viewed as a complex-wide reduction in fecundity. We begin our discussion of the discrete reaction–diffusion model by exploring its characteristics in chapter 6. In chapter 7 we present an example of the basic optimization model, borrowed from our previous book (Hof and Bevers 1998). In chapters 8–10 we investigate variations of the discrete reaction–diffusion model.

6

CHARACTERISTICS OF THE DISCRETE
REACTION–DIFFUSION MODEL

Before we discuss optimization models based on the discrete reaction–diffusion model developed in the introduction to part II, we explore the characteristics and behavior of this model (using simulation) and compare them with the reaction–diffusion theory literature. We use the cellular landscape definition and explore the effects of spatial habitat structure on population persistence, abundance, and distribution using numerical analysis of the implied "coupled map lattices" (Kaneko 1993). Using this approach to model spatial structure allows us to examine both within-patch and between-patch population dynamics with a single discrete time, discrete space reaction–diffusion model. The two-dimensional habitat complexes depicted in figure 6.1, for example, can all be modeled using cell-based reaction–diffusion as long as some degree of diffusivity exists among the cells.

In this chapter, we report some experiments from Bevers and Flather (1999b) in which populations were modeled that exhibit periodic reproduction and diffusion with adults and juveniles dispersing identically as

$$v_{it} = \sum_{j=1}^{N} \left[1 + f_j(v_{j(t-1)})\right] v_{j(t-1)} g_{ji} \quad \forall i, t, \tag{6.1}$$

where i and j each index all cells in the complex, v_{it} is the population in cell i at time t, $f_j(v_{j(t-1)})$ is the net per capita rate of reproduction within cell j (not account-

This chapter was adapted from M. Bevers and C. H. Flather, Numerically exploring habitat fragmentation effects on populations using cell-based coupled map lattices, *Theoretical Population Biology* 55 (1999): 61–76, with permission from the publisher, Academic Press.

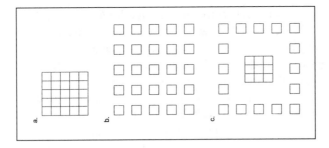

FIGURE 6.1

Three examples showing (a) single-patch, (b) hyperdispersed, and
(c) mainland–island habitat complexes consisting of 25 habitat cells.

ing for mortality associated with dispersal), and g_{ji} is defined as the proportion
of organisms in cell j expected to emigrate to cell i per time step. Equation (6.1)
is identical to equation (II.4), applied to the cellular landscape definition.

Experiments

In the discussion that follows, we use equation (6.1) to first investigate the
effects of patch size, shape, and heterogeneity on populations occupying a single patch. This is followed by a series of experiments on fragmented habitat
complexes. As noted before, fragmented landscapes often occur as habitable
areas embedded in a traversable but largely inhospitable matrix. We use
square habitat cells in this chapter, but other shapes (e.g., hexagons, as in chapter 5) could also be used.

The choice of spatial and temporal units requires scale consistency in the
reaction–diffusion process. For example, if the cells are breeding territories,
each g_{ji} diffusion parameter should reflect the probability of emigrating from
territory j to territory i a full breeding season later by any of a multitude of
possible routes. For simplicity, we assume identical diffusion from each cell
in any given complex, calculated in the following manner. Organisms disperse
on average from the center of each cell outward in uniformly random directions. Dispersal probabilities decline with distance (x) according to

$$p_X(x) = \mu^{-1} \exp\left[-(x - \theta)/\mu\right], \quad x > \theta, \ \mu > 0,$$

as defined by a mean dispersal distance (μ) from the center of the home cell
using a minimum dispersal distance (θ) of zero. Diffusion proportions (g_{ji}) are

estimated by numerical approximation over distances and angles defined by the boundaries of each destination cell (indexed by i) relative to the center of each source cell (indexed by j). Using an exponential distribution for dispersal distance (as in Fahrig 1992) results in a globally connected cellular lattice.

To observe the effects of within-patch heterogeneity clearly and simply, we hold reproduction constant by setting $f_j(v_{j,t-1})$ to a per capita net rate of reproduction r identical for every cell and every time period. In the absence of dispersal, this produces discrete exponential growth up to a constant adult population carrying capacity limit b_i that can be varied across cells by adding the constraint

$$v_{it} \leq b_i \quad \forall i, t. \tag{6.2}$$

In each time step, cellular populations from equation (6.1) in excess of equation (6.2) are treated as mortality resulting from dispersal into fully occupied territories. This procedure produces a probability of dispersal mortality (or reduced fecundity) across a varied landscape that is proportional to the amount of nonhabitat and fully occupied habitat around a given cell, with distance decay.

Populations Isolated in a Single Patch

Effects of Patch Size

Using continuous-variable random walk models, Skellam (1951) and others have demonstrated that for a single patch of habitat a critical patch size generally exists below which theoretical populations are expected to perish. In our examples, this threshold is defined by the combination of patch size, intrinsic rate of population growth (r), and mean dispersal distance (μ, hence all g_{ji}). By varying one parameter while holding the others constant, we can determine a critical set of threshold values. Below the threshold point, total population in the model declines toward zero from any initial population level.

To verify consistency of our model with reaction–diffusion theory, we demonstrate this critical threshold effect by considering a species with a mean dispersal distance (μ) of 2.50 cell side units, an r value of .495, and a carrying capacity (b_i) of 10 organisms per cell (as may be appropriate for social predators or cooperative breeders). By incrementally increasing the size of square patches, we initially observe a persistent population with a 7×7 cell patch (figure 6.2). This patch is at the critical size threshold for the given reaction–diffusion conditions, meaning that removing any cell from the patch results in an equilibrium population of zero. Similarly, if mean dispersal is

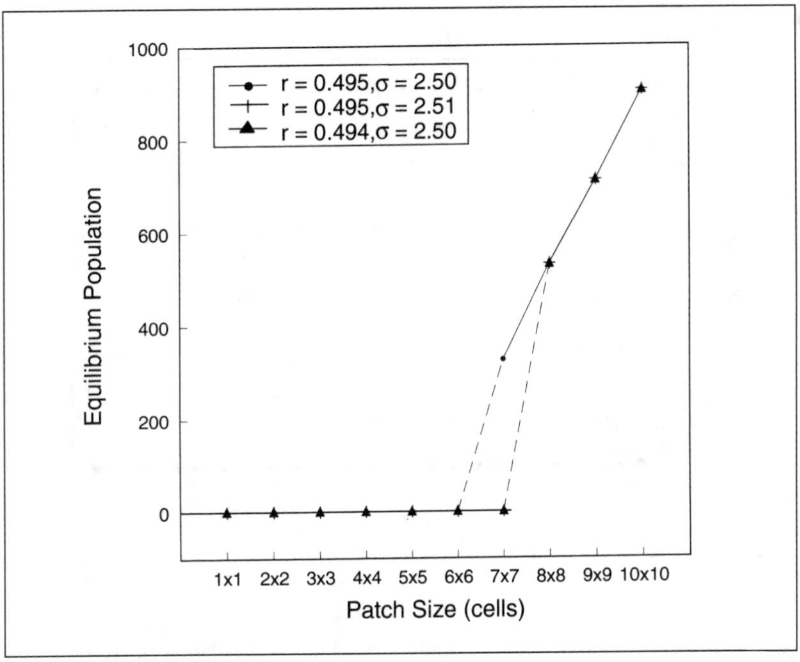

FIGURE 6.2

Equilibrium populations in square patches ranging in size from 1 to 100 cells
under three slightly different sets of reaction–diffusion parameter values.

increased to 2.51 (with $r = .495$) or if the r value is decreased to .494 (with
$\mu = 2.50$), the equilibrium population drops to zero.

Patch size also affects population growth and resulting density. Population
growth curves for square patches of different sizes under otherwise identical
conditions are shown in figure 6.3. Initial populations of 10 organisms per cell
were placed in the nine central cells of each patch so that population declines
would be visible for patches below the critical size threshold.

For the 7×7 cell patch near critical threshold conditions, we observe a *j*-
shaped exponential growth curve up to 327.5 population units, at which point
the population behaves as though the carrying capacity limit has been reached.
Given the constant r value used in this model, the shape of the growth curve
for this patch is not surprising except that the equilibrium population is only
about 66% of the total expected carrying capacity. It is also noteworthy that
about 2000 time steps are needed before the resulting upper limit is reached.
Consequently, the overall realized per capita rate of population growth is

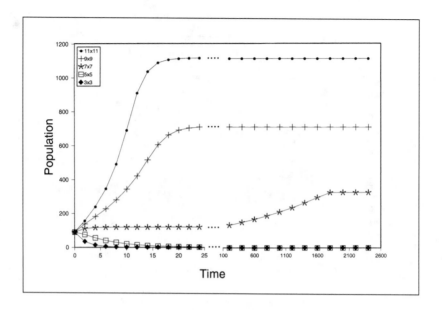

FIGURE 6.3

Population growth curves from centrally located initial populations of 90 organisms for five sizes of square habitat patches with reaction–diffusion parameters $r = .495$ and $\mu = 2.50$.

about .00065 per time step, much less than the r value used in the simulation. This realized rate of growth approaches zero as the critical extinction threshold is approached with greater precision. Below equilibrium in this system, dispersal mortality results entirely from diffusion into nonhabitat. Cellular capacity becomes limiting only at equilibrium.

Three other effects of patch size are particularly noticeable in figure 6.3. First, as patch size increases above the critical size threshold of 7×7 cells, the patch population growth curves become sigmoid rather than exponential in shape, despite constant net rates of within-cell reproduction, as dispersal mortality results more and more from movement into fully occupied cells. Second, the overall realized rate of population growth increases with patch size, although the realized rate is always less than the parameterized r value because of the loss of organisms dispersing beyond the patch and perishing. Third, as patch size increases, the resulting equilibrium population more closely approaches total expected patch carrying capacity, reaching about 88% of total carrying capacity in the 9×9 cell patch and about 92% of carrying capacity in the 11×11 cell patch.

A final patch size effect concerns the distribution of populations within the patch. Our results indicate that cellular populations are more variable, with systematic reductions toward the patch boundary, as critical patch size is approached (figure 6.4a). As patch size increases, the addition of centrally located cells results in a more uniform pattern of cellular populations (figure 6.4b). All these patch size effects are consistent with existing reaction–diffusion theory (see the review by Holmes et al. 1994).

Effects of Patch Shape

When dispersal is equally likely in all directions, Game (1980) and others suggest that circular shapes are preferred for single patches of remnant habitat. Ludwig et al. (1979) confirm that critical patch area is smallest for circular areas under such assumptions. Using uniformly distributed dispersal as we do here, we would expect elongations in patch shape to affect overall population growth and persistence in our model. To confirm this, we consider three patches, each comprising 144 cells 1 unit in size but arranged with increasing elongation. The curves in figure 6.5 result from a centrally located initial population of 1 unit (with $r = .495$ and $\mu = 2.50$). As the shape elongates from a square to the 4:1 length-to-width ratio, early growth lags, the slope of the growth curve declines, and the upper population limit drops. When the shape is elongated further (16:1), the patch drops below a critical shape threshold and cannot support any long-term population. This result again conforms with existing reaction-diffusion theory (see Holmes et al. 1994).

Effects of Intrapatch Heterogeneity

Intrapatch heterogeneity can be introduced into the model by varying the cellular carrying capacity limits (b_i), allowing us to begin exploring habitat configurations that have proved difficult to address with earlier models. The individual effects of variations in cellular capacity depend on that cell's location within the modeled complex and on the associated reaction–diffusion system. In a 7×7 cell patch with $r = .495$ and $\mu = 2.50$, setting the carrying capacity of the center cell to 1, with all of the others set to 10, results in an equilibrium population of 32.75 distributed as in figure 6.6a. Because the patch is very close to threshold conditions, a 90% lower capacity in just the center cell has the same effect as a 90% reduction in all cells. Although a 90% reduction in carrying capacity within a patch at threshold conditions does not drive the expected population to zero in this deterministic model (unlike the removal

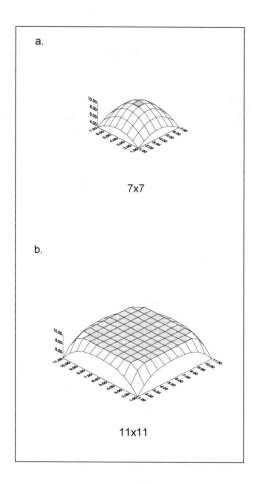

FIGURE 6.4

Cellular populations at equilibrium for (a) a 7×7 and (b) an 11×11 cell patch with reaction–diffusion parameters $r = .495$ and $\mu = 2.50$ and centrally located initial populations of 90 organisms. Shading indicates cells at carrying capacity.

of a single cell), the effects are large and the resulting population would be much more likely to perish in a stochastic simulation (Goodman 1987). Corner cells in a square patch are the least connected in a diffusion sense, so lowering the capacity of a corner cell has less effect, resulting in an equilibrium population of 114 (shown in figure 6.6b). Near threshold conditions, the locations of habitat quality changes have a significant effect on total abundance within the patch.

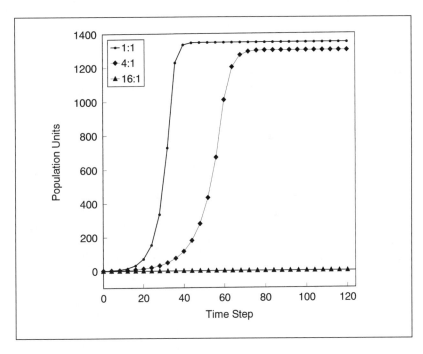

FIGURE 6.5

Population growth curves from centrally located initial populations of 1 unit with reaction–diffusion parameters $r = .495$ and $\mu = 2.50$ for 144-cell rectangular patches of three different shapes, indicated by length-to-width ratio.

Populations Occupying Multiple Patches

One of the principal strengths of this model is that similarities and differences in population responses to both contiguous and fragmented habitat complexes can be examined without any structural change in the formulation. In the following experiments, we examine these similarities and differences by investigating the effects of habitat fragmentation.

Fragmentation of Contiguous Habitat

Allen (1987) shows the existence of a critical number of patches in a reaction–diffusion model of one-dimensional, locally coupled patchy habitats with only end point dispersal into nonhabitat. This suggests that we should be

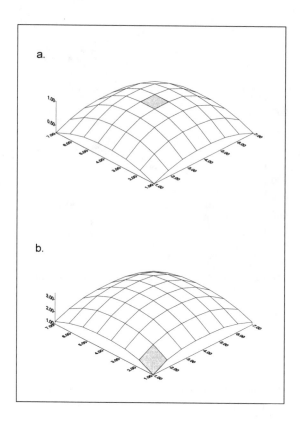

FIGURE 6.6

Cellular populations at equilibrium in a 7×7 cell patch with reaction–diffusion parameters $r = .495$, $\mu = 2.50$, and carrying capacities of 10 population units in all cells except (a) the center cell and (b) a corner cell (shaded), for which carrying capacities are set to 1.

able to observe persistence thresholds for fragmented habitat complexes similar to those we have observed for contiguous patches.

Earlier, we described a 7×7 cell patch that slightly exceeded critical habitat threshold conditions for a population with $r = .495$ and $\mu = 2.50$. With carrying capacity set at 10 population units per cell, the equilibrium patch population was 327.5. If we fragment that contiguous block of cells into 49 one-cell patches arranged in a hyperdispersed fashion (as in figure 6.1b), dispersal-related mortality increases markedly. Consequently, the habitat complex can no longer support a persistent population under those reaction–diffusion conditions. Indeed, population persistence is not observed until the

r value is increased to 2.172 (resulting in an equilibrium population of 296.2) or until mean dispersal is reduced below .53 (implying that almost two-thirds of the organisms remain in their home cell from time period to time period, resulting in an equilibrium population of 312.2). Because none of the patch populations in the complex are individually persistent, we would expect such a complex to function as a metapopulation (Harrison and Taylor 1997).

Consistent with the notion of a critical number of patches (Allen 1987), the population fails to persist at $r = 2.171$ and $\mu = 2.50$ unless more habitat is added to the complex. With the addition of a single similarly displaced one-cell patch off the lower right-hand corner of the complex, a persistent equilibrium population of 293.3 organisms results. Thus, for a population occupying hyperdispersed one-cell patches in a square arrangement, with the reaction–diffusion parameters just given, the critical number of patches is 50.

As we increase the overall size of the hyperdispersed habitat complex from 7×7 to 9×9 and 11×11 one-cell patches (with $r = 2.172$ and $\mu = 2.50$), the population growth curves shift, as in the single- patch case (figure 6.3), from j-shaped to sigmoid (figure 6.7). However, the rate of population growth is slower than in the contiguous habitats portrayed in figure 6.3 (from similar initial populations). Equilibrium populations approach total available carrying capacity as the habitat complex is expanded beyond threshold conditions, although again at a slower rate than observed with contiguous habitats.

Pronounced differences in critical threshold effects between contiguous and patchy habitats can also be observed, however. For example, after noting that a four- to fivefold increase in r value is enough to largely compensate for the effects of fragmenting our 7×7 cell habitat complex, we might expect that (with $r = .495$ and $\mu = 2.50$) population recovery could be achieved by providing a four- to fivefold increase in the size of the fragmented complex. In our experiments, though, even a 200×200 hyperdispersed complex of one-cell patches was inadequate to support a persistent population in these circumstances.

On one hand, we have a set of results that support Allen's (1987) finding that for a given r value and dispersal rate, it is always possible to reach persistent equilibrium population conditions by adding enough similarly configured habitat. On the other hand, we have a set of results suggesting instead that because the ratio of habitable cells to uninhabitable cells and their geometry remain unchanged, no amount of additional similarly arranged habitat would be enough. This dilemma is resolved by examining the dispersal parameters.

In our 7×7 hyperdispersed complex, the center patch is best positioned to supply successful dispersers with habitat, given our dispersal assumptions. From that patch, organisms have a probability of about .34 of dispersing into any one of the 49 habitat patches in the complex. They have a probability of

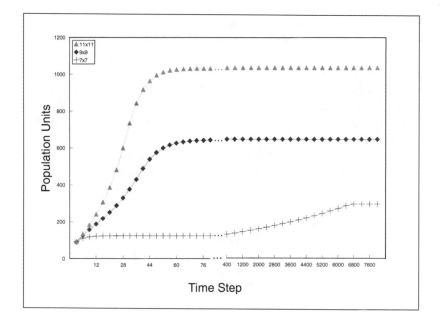

FIGURE 6.7

Population growth curves from centrally located initial populations of 90 organisms with reaction–diffusion parameters $r = .495$ and $\mu = 2.50$ for varying sizes of habitat complexes fragmented into hyperdispersed arrangements of one-cell patches.

about .60 of dispersing into nonhabitat areas within the complex perimeter and perishing and a probability of about .06 of dispersing beyond the perimeter of the complex and perishing. If all 49 cells had similar circumstances, the population would need an r value of about 1.94 to persist in this model. We estimate this by observing that $(1 + r)$ times the fraction that will disperse successfully (.34) has to equal or exceed 1 (implying no net population loss) for the population to persist in our model.

Because the other 48 patches are actually less well positioned than the center patch (and have greater dispersal mortality), an r value slightly greater than 2.17 is needed. Furthermore, adding similarly configured habitat patches around the complex (i.e., one of every four cells is habitat; figure 6.1b) can only increase the probability of successful dispersal from the center cell by about one-fourth of .06, suggesting a minimum r value for persistence somewhat greater than 1.82. We calculate this from $[(.34 + .06/4)(1 + r) = 1]$ by estimating that if hyperdispersed one-cell habitat patches were added throughout the region extending beyond the present complex perimeter, about one of every four organisms would be able to disperse successfully from the center cell. Consequently, we can expect some possibility of reaching persistence by adding hyperdispersed one-cell patches for populations with r values between roughly 1.82 and 2.17 given our dispersal assumptions. Below an r value of about 1.82, no amount of additional similarly configured habitat will suffice without more restrictive dispersal assumptions.

Island Systems

In a mainland–island system connected by population diffusion, with the island being too small to independently support a persistent population, we generally expect the island population to have little influence on the size of the overall population. In an archipelago system with no mainland, on the other hand, we generally expect each island to have a stronger influence on overall population size (again assuming that each island is too small to independently support a persistent population). Stepping-stone systems arise because in many cases a practical limit to dispersal distance exists such that organisms are unable to move directly between all patches in the habitat complex, forming an incompletely coupled lattice.

One of the more interesting questions regarding population persistence in these insular systems concerns the effect of diffusion parameters on the relative importance of small, isolated patches. We investigate this question in mainland–island and archipelago systems using three separate habitat patches, arranged diagonally and decreasing in size (figure 6.8). If this complex is inhabited by a species with a mean dispersal distance of 2.00, an r value of .571, and a carrying capacity of 10 organisms per cell, the complex just exceeds critical threshold conditions for supporting a persistent population of 173.0 organisms. However, the two smaller islands are not essential for persistence. The 5×5 patch alone exhibits an equilibrium population (170.6 organisms). Because a mean dispersal distance of 2.00 units results in little interpatch diffusion, this complex functions as a mainland–island system with very small expected populations on the two islands, which would be expected to "wink" (Gilpin 1987) in and out of existence in a stochastic simulation with the population modeled as discrete organisms.

For species that exhibit greater mean dispersal distances, a true archipelago system emerges. Consider a species with a mean dispersal distance of 4.00 units. In this habitat complex, such a species would need an r value of 1.351 just to achieve critical threshold conditions for persistence. A carrying capacity of 10 organisms per cell would then produce an equilibrium population of 191.7 organisms. For this species, however, the 5×5 cell patch is no longer

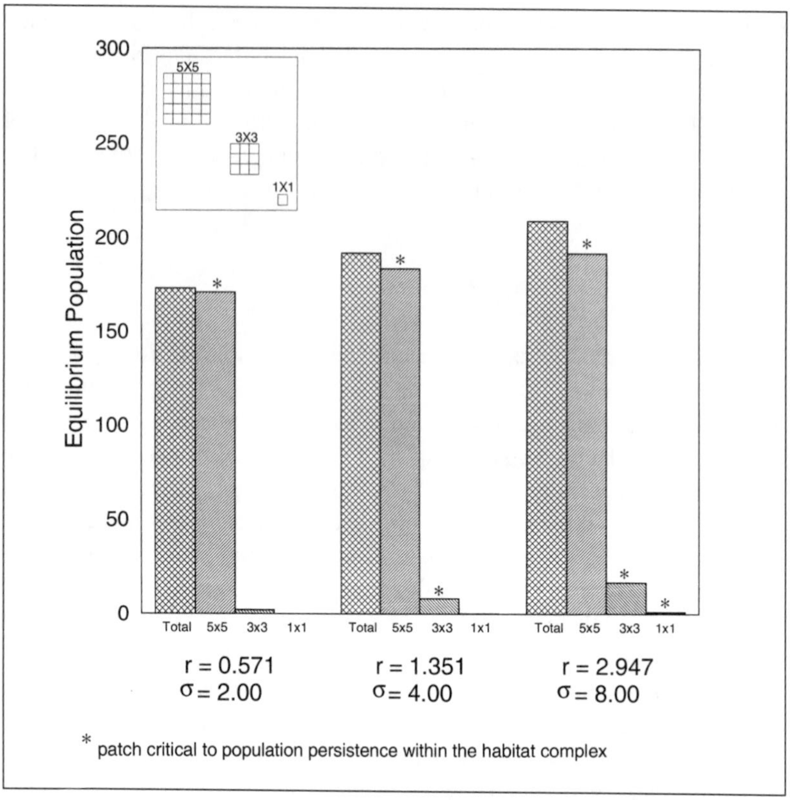

FIGURE 6.8

Total and patch-level equilibrium populations and critical patches in a three-patch complex with a total expected carrying capacity of 350 organisms (see inset) under three sets of threshold-level reaction–diffusion parameters reflecting increasing mean dispersal and net reproduction.

able to independently support persistence; instead, both larger islands are needed (the smallest island remains unessential and barely populated). At a mean dispersal distance of 8.00 units (and an increase to $r = 2.947$ to meet critical threshold conditions), all three islands become essential to support a persistent equilibrium population of 208.7 organisms. Near persistence thresholds for this habitat chain, a shift toward metapopulation conditions occurs as mean dispersal distance increases to encompass most or all of the complex.

Under passive diffusion assumptions, high reproduction rates are needed to support such metapopulations, and only slightly higher rates cause a shift back to a mainland–island system. At a mean dispersal distance of 4.00 units, increasing the r value from 1.351 (the persistence threshold as a metapopulation) to 1.355 allows the population on the largest island to persist independently. At a mean dispersal distance of 8.00 units, increasing the r value from 2.947 to 2.977 causes similar results. These narrow ranges in r values and their proximity to extinction thresholds suggest that classic metapopulation dynamics (Harrison and Taylor 1997) would occur only rarely in passively diffusing populations.

Using the exponential distribution for dispersal distances, as we do here, unrealistically portrays the diffusion process as occurring over infinite distances in each time step. Instead, organisms generally are limited to some finite maximum dispersal distance over a fixed time interval. This leads to the possibility that small stepping-stone habitat patches might be used to provide diffusion between two larger but otherwise isolated habitat complexes.

To investigate the effects of stepping-stone patches, we constructed a model with two 5×5 cell habitat patches placed edge-toward-edge at a distance of three cells apart. Centered between these two patches, we placed a one-cell stepping-stone patch (see figure inset in table 6.1). Diffusion parameters were calculated as before, using an exponential distribution with a mean dispersal distance of 1.50 cell-side units but truncated at a maximum distance of 3.50 units. Because diffusion coefficients are estimated from the center of each source cell in our numerical analysis, no diffusion occurs directly between the two larger patches.

With an r value of .401 and carrying capacities set to 10 organisms per cell, the two large patches can each independently support an equilibrium population of 161.3 organisms at threshold conditions (table 6.1). Adding the stepping-stone patch increases the total complex population from 322.6 to 330.2 organisms, with 1.9 of those in the stepping-stone patch. Without the stepping-stone, reducing the r value to .400 drops both of the larger patches below critical threshold conditions. With the stepping-stone in place, an r value of .400 merely reduces the equilibrium population to 326.0 organisms.

Although this demonstrates that stepping-stone patches can support persistence of modeled metapopulations, the usefulness of such patches can be very limited. A slight reduction in net reproduction from .400 to .398 causes the metapopulation to perish even with the stepping-stone patch. Thus the range of r values for which this well-placed stepping-stone patch supports persistence is only from .399 to .400. Similarly, if we increase the distance between the two larger patches from 3 to 5 units (with $r = .400$) and again place the stepping stone halfway between the larger patches, the metapopulation perishes. These results suggest that successful use of stepping-stone

TABLE 6.1

Equilibrium Populations in Stepping-Stone Island Systems
with Varying Population Growth Rates

Habitat Layout		*Population Growth*			
		$r = .401$	$r = .400$	$r = .399$	$r = .398$
	Total without stepping stone	322.6	0	0	0
	Total with stepping stone	330.2	326	321.9	0
	Stepping stone	1.9	1.9	1.8	0

All simulations were run with mean dispersal distance (μ) set to 1.5 units, with a maximum dispersal distance of 3.5 units. The habitat layout is shown in the figure inset.

patches in population recovery strategies would be improbable under our dispersal assumptions.

Effects of Patch Shape on Colonization

Diamond and May (1976), Game (1980), and others have suggested that new patches introduced into existing habitat complexes would be colonized more quickly given an elongated shape rather than circular or square shapes. However, Hamazaki (1996) points out that elongated perimeters could enhance emigration as much as or more than immigration, confounding the expected results. With equal-size cells and equivalent diffusion for all cells, our examples retain the property that cell-to-cell diffusion is identical for both emigrating and immigrating organisms. Consequently, we can use our model to examine the effects of patch shape on colonization in the absence of immigration or emigration bias by placing unpopulated patches identical in size but of different shapes within a populated habitat complex.

We constructed a habitat complex of 140 cells connected together to approximate the circumference of a circle about 36 cell-sides in diameter to represent the surrounding existing habitat for this analysis. With $r = 1.75$, $\mu = 2.50$, and capacities set to 10 organisms per cell, this ring of habitat supports an equilibrium population of 1333 organisms. We used this population as our initial condition.

We placed three different shapes of unpopulated habitat 64 cells in size into the center of the existing habitat ring and modeled population growth (figure 6.9). For the first several time steps, the most elongated patch (a 2 × 32 cell rectangle) clearly demonstrates a more rapid net gain in population. The moderately elongated patch (a 4 × 16 rectangle), on the other hand, shows somewhat slower population growth than the square patch, although both show more rapid gains than the 2 × 32 cell patch shortly before approaching equilibrium conditions. Finally, the most elongated patch appears to support a slightly greater long-term population than the less elongated patches. However, nearly all the additional long-term population actually occurs in the surrounding ring, probably because the 2 × 32 cell patch places several cells much closer to the surrounding ring than do the other two patches.

To test whether these effects of patch shape are sensitive to the reaction–diffusion parameters, similar analyses were performed with $r = 2.25$ and $\mu = 2.50$ and 3.50. All results were similar to those shown in figure 6.9. Elongated new patches in existing habitat complexes appear to consistently show some early colonization advantages under passive diffusion assumptions, whereas blockier patches tend to support faster population growth later on.

Whether newly colonized elongated or blocky patches will support higher overall equilibrium populations probably depends on the arrangement of the surrounding habitat and on mean dispersal distance. Moreover, once the new patch becomes necessary for persistence, serving as the mainland of the habitat system, our square patch can support population increases in the complex as mean dispersal distance increases, whereas the better-connected elongated patches cannot (table 6.2). Although these increases are slight, they suggest that the addition of a blockier critical habitat patch might increase resilience to variations in dispersal distance.

Discussion

In this chapter, we confirm that results from our discrete model conform to those generally expected from reaction–diffusion models, as summarized by Holmes et al. (1994). The single-patch effects include the existence of critical patch size thresholds (figure 6.2), increases in overall population growth rate and more fully occupied habitat as patch size increases above the critical size threshold (figure 6.3), Gaussian-like equilibrium population distributions in uniform habitat (figure 6.4), and the observation that critical patch size is smaller for square patches than for elongated shapes (figure 6.5). Furthermore, we confirm the prediction that increasing dispersal out of the patch leads to larger critical patch sizes. In our model, increases in mean dispersal distance

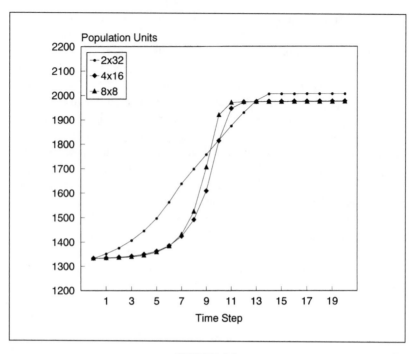

FIGURE 6.9

Population growth curves beginning from equilibrium conditions in an existing habitat complex with reaction–diffusion parameters $r = 1.75$ and $\mu = 2.50$ after the introduction of an unoccupied, centrally placed 64-cell patch of habitat varying in shape.

have this effect and must be compensated for by increases in r value or patch size for populations to be maintained. These outcomes are consistent with those previously reported from reaction–diffusion studies and provide a reasonable basis for experiments with heterogenous habitat patches and fragmented patch complexes.

In nonuniform single-patch systems, our experiments on the effects of heterogenous habitat capacities demonstrate that the location of habitat quality changes become more important near extinction threshold conditions. At critical patch size, equilibrium populations appear to approach Gaussian-like spatial distributions of abundance, with total population constrained by the capacity of the most limiting (in a reaction–diffusion sense) cell (figure 6.6).

Similar effects can be shown for populations in fragmented habitat landscapes using otherwise identical models (e.g., figure 6.7). However, our experiments clearly demonstrate that these systems have inherently more complex behavior that is parametrically sensitive. For example, the existence

TABLE 6.2

Equilibrium Populations at Various Mean Dispersal Distances (μ)
for Isolated and Combined Habitats of Different Configuration at $r = 1.75$
with Cellular Capacities Set to 10

μ	140-Cell Ring	8 × 8 Patch	Ring with 8 × 8 Patch	4 × 16 Patch	Ring with 4 × 16 Patch	2 × 32 Patch	Ring with 2 × 32 Patch
2.5	1332.9	640	1976	640	1978.3	640	2007
3.5	0	640	676.3	640	688.1	614.3	818.9
4.5	0	638.5	678.8	631.8	679.8	0	0
5.5	0	633	680.6	607.2	659.4	0	0
6.5	0	613.4	666.2	516.5	568.4	0	0
7.5	0	555	608.5	0	0	0	0
8.5	0	0	0	0	0	0	0

of a critical size threshold in single-patch systems guarantees that population persistence can be achieved by increasing the size of the patch regardless of the magnitude of the other system parameters. For hyperdispersed patchy habitats, on the other hand, we show that the ability to achieve persistence by extending the size of the complex (i.e., providing more similarly fragmented habitat) depends on the reaction–diffusion parameters in the system. At sufficiently low r values, such an approach cannot succeed in our model regardless of the increase in size of the complex. Because these systems could be made persistent by creating a single individually persistent patch, the population recovery efforts that are modeled in chapters 7–10 often focus on strategies that increase existing patch sizes when those options are available.

We see parametric sensitivity in our results for other fragmented landscapes, as well. For an island chain habitat complex, metapopulation conditions in which no single patch is individually persistent arise only when both dispersal distances and reproduction rates are relatively high (figure 6.8) and only within very narrow parameter ranges that are close to extinction thresholds. These results offer one explanation for Harrison and Taylor's (1997) finding that classic metapopulations appear to be scarce. In terms of population distribution, persistent archipelago complexes do not appear to be substantially different from mainland–island systems. Empirically, they may be indistinguishable, and stochastically, extinctions complexwide may be nearly as frequent as extinctions from large islands.

Similarly, our results suggest that successful use of stepping-stone patches to support metapopulation systems would be unlikely for passively diffusing species (table 6.1). Once again, these results tend to suggest that the models in chapters 7–10 will tend to increase habitat patch sizes and provide block-ier patch shapes within existing fragmented landscapes to achieve population recovery. This is consistent with Adler and Nuernberger's (1994) finding that clustering identical patches in a fragmented habitat complex enhances popu-lation persistence. Although elongated patches added to existing complexes do offer early colonization advantages in our model (figure 6.9), blocky patches appear to offer greater long-term population resilience to fluctuations in reaction–diffusion behavior near threshold conditions (table 6.2).

Our models are based on simple assumptions about population behavior. Passive diffusion identical for adults and juveniles, a completely uninhabit-able matrix between habitat patches, and constant per capita reproduction are fairly strong assumptions, even though similarly simple assumptions have been used successfully to model populations of many species, including humans, with reaction–diffusion equations (e.g., Ammerman and Cavalli-Sforza 1984). The effects of relaxing these assumptions in reaction–diffusion models generally are well understood for single-patch systems (see Holmes et al. 1994:22–23). For example, factors that decrease diffusion from patches generally lead to smaller critical patch sizes. Such factors include advective diffusion toward available habitat, density-dependent dispersal, and nest site fidelity. Likewise, less hostile boundary conditions generally lead to smaller critical patch sizes.

Capturing more realistic dispersal behavior by adding age structure as we discuss in the part II introduction is tempting but entails a substantial increase in data that is difficult to obtain and is highly variant. Wennergren et al. (1995) question the usefulness of such added complexity. As noted in the part II intro-duction, increasing age complexity to separately represent adult and juvenile dispersal results in a model that differs only parametrically from the simpler model used for our experiments.

Density-dependent population growth rates tend to regulate population sizes but do not affect critical patch size in the absence of Allee effects (Mur-ray 1989). Furthermore, population growth occurs within breeding territories in our model. Although equation (6.1) can incorporate density-dependent growth effects, those effects are modeled in our experiments at the population scale as a consequence of dispersal-related mortality (which could also rep-resent an equivalent reduction in fecundity). This is consistent with the notion that mortality during dispersal can have a stabilizing influence on populations (Ruxton et al. 1997). Unlike models in which dispersal-related mortality is not apparent (e.g., Levin 1974; Hastings 1982, 1992; Bascompte and Solé 1994;

Hassell et al. 1995), it is portrayed explicitly in our model through the cumulative dispersal parameters.

One final implication for our models, based on observed density declines toward habitat edges in uniform patches, is that population density reductions at habitat perimeters among species considered to be habitat interior specialists may be partially explained by random dispersal patterns rather than selection for interior habitats. Similarly, in chapters 7–10, slow population growth and spread rates during early observations of recovering species could result from poorly suited habitat, Allee effects, or a combination of random diffusion, habitat configuration, and initial population placement.

7

THE BASIC MODEL: HABITAT PLACEMENT
FOR THE BLACK-FOOTED FERRET

The Black-Footed Ferret

Early in 1987 the black-footed ferret (*Mustela nigripes*) became one of the world's most endangered mammals when the last known free-ranging member of the species was taken into captivity (Thorne and Belitsky 1989). The Wyoming Game and Fish Department was successful in breeding six of the surviving ferrets in captivity (Clark 1989), setting the stage for a national recovery program of releasing captive-bred ferrets back into the wild.

Historically, the black-footed ferret ranged sympatrically with prairie dogs (*Cynomys* spp.) across much of North America (Anderson et al. 1986). Available evidence strongly supports the conclusion by Henderson et al. (1969) that black-footed ferrets have narrow habitat requirements, living principally in prairie dog burrows and depending primarily on prairie dogs for prey (Linder et al. 1972). Demise of the species in the wild has been attributed to loss and fragmentation of habitat (prairie dog colonies) caused by extensive prairie dog eradication programs and changes in land use, combined with susceptibility of prairie dogs to sylvatic plague and of ferrets to canine distemper (U.S. Fish and Wildlife Service et al. 1994). As Seal (1989) points out, it now appears difficult to find suitable ferret habitat complexes ("groups of prairie dog colonies

This chapter was adapted from M. Bevers, J. Hof, D. Uresk, and G. Schenbeck, Spatial optimization of prairie dog colonies for black-footed ferret recovery, *Operations Research* 45, no. 4 (1997): 495–507, with permission from the publisher, the Institute for Operations Research and the Management Sciences (INFORMS).

in close proximity," Biggins et al. 1993) of 3000 to 15,000 hectares, even though prairie dogs were once distributed over millions of hectares of land.

The first release of captive-bred black-footed ferrets into the wild occurred in 1991 in Shirley Basin, Wyoming. Two additional reintroduction areas were added in 1994, including the site of this study centered in Badlands National Park, South Dakota. These ferret release sites were selected on the basis of habitat suitability and other biological and sociopolitical factors.

The spatial arrangement of prairie dog colonies in a colony complex has important effects on the number of black-footed ferrets that can be supported (Minta and Clark 1989). As prairie dog colonies become smaller or more widely separated, successful ferret dispersal between colonies is less likely, and the total population that can be supported is reduced. Houston et al. (1986) and Miller et al. (1988) have used spatial measures such as mean intercolony distance and colony size frequency distribution in estimating ferret habitat suitability, but Biggins et al. (1993) note a number of troubling quantitative difficulties with such approaches. For example, it is often possible to identify a number of habitat patch arrangements that are equal in mean intercolony distance (as well as individual and total patch sizes) for which expected population responses typically would not be equal according to biodiffusion theory (Okubo 1980). Consequently, in the Biggins et al. procedure, the effects of spatial colony arrangement within colony complexes are assessed qualitatively.

This chapter presents the more rigorous quantitative approach derived in the part II introduction and explored in chapter 6. The model described in this chapter was used to analyze habitat management and ferret release as a spatial efficiency problem on the federally managed lands of the Buffalo Gap National Grassland adjacent to the Badlands National Park ferret release area.

The Model

Within the reintroduction area, black-footed ferret habitat comprises a complex of active and potential prairie dog colonies (patches) forming distinct habitat islands on the landscape. Our model used discrete annual time periods and approximated the habitat spatial configuration with a grid of cells on the landscape. We then incorporated cellular habitat management decision variables so that we could consider all potential spatial configurations (within the resolution of our grid). Particular ferret habitat layouts in the model are achieved over time by the prairie dog populations that result from the rodenticide treatment–nontreatment schedules applied to each cell across the landscape, on the premise that prairie dog populations recover rapidly in areas left untreated.

Adult black-footed ferret populations in each cell in any year are limited in the model by either the carrying capacity of that cell, the ability of ferrets from nearby cells to successfully reproduce and disperse there, or both. Additional decision variables (R_{it}) were used to determine the timing (year t) and location (cell i) for captive-bred ferret releases into the area. Our spatial optimization model is as follows:

Maximize

$$F_T, \tag{7.1}$$

subject to

$$F_t = \sum_i S_{it} \qquad t = 1, \ldots, T, \tag{7.2}$$

$$S_{i0} = N_i \quad \forall i, \tag{7.3}$$

$$S_{it} \leq R_{it} + \sum_j g_{ji}(1 + r_j)S_{j(t-1)} \quad \forall i \qquad t = 1, \ldots, T, \tag{7.4}$$

$$\sum_i R_{it} \leq b_t \qquad t = 1, \ldots, T, \tag{7.5}$$

$$S_{it} \leq \sum_{h=1}^{H_i} \sum_{K=1}^{K_{ih}} c_{ihkt} X_{ihk} \quad \forall i \qquad t = 1, \ldots, T, \tag{7.6}$$

$$\sum_{k=1}^{K_{ih}} X_{ihk} = A_{ih} \quad \forall i, h, \tag{7.7}$$

$$\sum_i \sum_{h=1}^{H_i} \sum_{k=1}^{K_{ih}} c_{ihktp} X_{ihk} \leq C_{pt} \quad \forall p \qquad t = 1, \ldots, T. \tag{7.8}$$

Indexes

t indexes annual time periods $(0, \ldots, T)$ that begin in late spring when the young emerge from the den,
i, j indexes habitat management cells in the study area,
h indexes initial habitat condition classes in each cell,
k indexes rodenticide treatment schedules,
p indexes policy constraints on selection of habitat management variables.

Decision Variables

X_{ihk} = amount of area (in hectares) in cell i and initial habitat condition class h allocated to the kth multiyear habitat management (rodenticide treatment) schedule,

R_{it} = number of captive-bred ferrets released in cell i during year t that are expected to survive adaptation to life in the wild.

Ferret Population Variables

S_{it} = adult ferret population (including yearlings) in cell i at the beginning of year t, plus R_{it},

F_t = adult ferret population for the entire complex in year t.

Parameters

N_i = estimated initial number of adult ferrets in cell i,

g_{ji} = proportion of surviving adult and juvenile ferrets from den areas in cell j in year $t - 1$ expected to disperse and become adult ferrets in cell i at the beginning of year t,

r_j = an r value for ferrets in cell j reflecting the maximum expected annual net population growth rate (i.e., when habitat is not a limiting factor),

b_t = an upper bound on the total number of captive-bred ferrets released during year t expected to survive adaptation to life in the wild,

c_{ihkt} = adult black-footed ferret carrying capacity for cell i and condition class h in year t per hectare allocated to X_{ihk},

H_i = number of initial habitat condition classes (based on prairie dog population densities) in cell i,

K_{ih} = number of habitat management schedules being considered for initial habitat condition class h in cell i,

A_{ih} = total prairie dog colony area (in hectares) of initial habitat condition class h in cell i,

c_{ihktp} = equals c_{ihkt} if X_{ihk} could contribute to policy constraint p in time period t and zero otherwise,

C_{pt} = amount of total black-footed ferret carrying capacity (i.e., prairie dog population) allowed in time period t under policy p from the relevant subset of X_{ihk} habitat management variables.

Equations (7.1)–(7.5) define a discrete time reaction–diffusion system (in this case, population growth and dispersal) for evaluating population persistence within a habitat complex. Equation (7.1) maximizes total adult black-footed ferret population at the end of year T, as summed by equation (7.2). Equation (7.3) sets initial population conditions. Equation (7.4) limits the adult ferret population in any cell for any year to (at most) the number of captive-bred ferrets released into the cell plus the number of ferrets expected to disperse into the cell from all cells in the complex (including the same cell), after accounting for net reproduction during the previous year. Equation (7.5) limits the number of ferrets that can be released during any given year.

Reaction–diffusion models with areas of nonhabitat generally assume that organisms dispersing into unsuitable regions will perish. This mechanism provides a probabilistic basis for the expectation that after accounting for births and deaths from all causes in an abundant habitat setting through the r value (net annual population growth rate r_j), additional mortality will occur in proportion to the usage of inhospitable surroundings. Equations (7.6) and (7.7) account for these habitat dynamics by imposing black-footed ferret carrying capacity constraints in each cell as a function of the selected habitat management (rodenticide treatment) schedules. The expected ferret population in any cell in a given year (S_{it}) is determined by equation (7.4) or (7.6), whichever is limiting. Individual or combined cellular carrying capacities may also be limited by prairie dog population management policies that restrict the availability of ferret habitat through equation (7.8). We used equation (7.8) to limit the total amount of ferret habitat in the National Grassland to examine spatially and temporally efficient tradeoffs between expected adult ferret populations and levels of prairie dog population control.

The linear dispersal model (equation 7.4) is based on an assumption of purely random diffusion, which is a first-level approximation for a highly developed species such as the black-footed ferret. More realistic ferret dispersal patterns, if they were known, might exhibit biased diffusion, such as movement in response to overcrowding (Gurney and Nisbet 1975) or selective movement based on acquired knowledge of active prairie dog colony locations or surrounding terrain. In general, biased diffusion enhances the persistence of populations (Allen 1983). Thus our model provided an estimated lower bound on the size of the expected population (not accounting for stochastic effects).

In some cases exponential population growth within large cells may be unreasonably optimistic. For more conservative growth rates, sigmoid population growth, by cell, could be approximated in a linear model by replacing equation (7.4) with

$$Q_{jt} \le (1 + r_j)S_{j(t-1)} \qquad \forall j \qquad t = 1, \ldots, T,$$

$$Q_{jt} \le S_{j(t-1)} + a_j \qquad \forall j \qquad t = 1, \ldots, T,$$

$$S_{it} \le \sum_j g_{ji} Q_{jt} \qquad \forall i \qquad t = 1, \ldots, T,$$

where an accessory variable for cell population in each year (Q_{jt}) is limited to either exponential (r_j) or incremental (a_j) growth up to carrying capacity, depending on the magnitude of $S_{j(t-1)}$. Note that we identified the r_j and a_j growth rates by cell. In some cases, expected net reproduction may be different from one cell to another even with abundant habitat. For example, one cell may lie closer to terrain frequented by predators than another cell.

By limiting the total annual releases, as in equation (7.5), the spatial optimization model was used to help identify preferred ferret release locations. When release locations are predetermined, the annual release variables (R_{it}) can be individually limited instead of constraining the annual sums. We also assumed that surviving released ferrets would disperse and reproduce similarly to indigenous ferrets. Additional mortality typical of released ferrets during the establishment period was accounted for by setting the upper bounds on releases (b_t) to the number of released ferrets that were expected to survive and reproduce.

In cases such as black-footed ferret reintroductions, where initial conditions differ greatly from potential persistent population levels, habitat conversion strategies based on prairie dog population management decisions in the first several years can be as important as long-term management strategies. The relative weight placed on short-term versus long-term management depends primarily on the choice of objective function (e.g., Bevers et al. 1995). Although the objective function expressed in equation (7.1) is suitable for estimating persistent (long-term) expected ferret population levels, another objective function was also used in the case study to place more emphasis on early ferret establishment. We replaced equation (7.1) with the following:

Maximize

$$\sum_t F_t, \qquad (7.9)$$

for all analyses except those focused on long-term persistence.

Ferret Reintroduction in South Dakota

Between September 19 and November 14, 1994, 36 captive-bred black-footed ferrets were released near the center of the Sage Creek Wilderness in Badlands National Park (McDonald 1995). This was the first of five annual releases planned for the purpose of reestablishing ferrets in the National Park and the surrounding Buffalo Gap National Grassland (Plumb et al. 1994).

We applied the spatial optimization model to a region approximately 157,500 ha in size surrounding the South Dakota reintroduction area, assuming that prairie dog population controls would preclude ferret recovery outside this study area. Federally managed lands with active (supporting substantial live prairie dog population densities) or readily recoverable inactive black-tailed prairie dog (*C. ludovicianus*) colonies suitable for ferret habitat over the next 10–15 years were fragmented, occupying less than one-tenth of the study area. Badlands National Park contained an estimated 2403 ha of active prairie dog colonies, primarily in the rugged Sage Creek Wilderness. This acreage was not expected to change significantly over the 10- to 15-year planning horizon. Under management plans in place at the time, Buffalo Gap National Grassland supported an estimated 2112 ha of predominantly active prairie dog colonies reserved from rodenticide use adjacent to Badlands National Park. We refer to these as initial colonies for 1994, the beginning year of the model ($t = 0$). The grassland contained an additional estimated 9850 ha of predominantly inactive prairie dog colonies in the study area that had been treated with rodenticide in previous years. We refer to these as potential colonies.

Spatial Definition

We selected U.S. Public Land Survey sections as cells for the model and assumed for dispersal probability calculations that each of the 608 survey sections (indexed by i) in the study area was a square enclosing 259 ha of land. The area (in hectares) of existing prairie dog colonies within each section was estimated from color infrared aerial photography taken in August 1993 using methods described by Schenbeck and Myhre (1986) and Uresk and Schenbeck (1987). Active prairie dog colony areas within these intact burrow systems were inventoried from field survey records. Inactive areas were identified as those that were readily recoverable in the 10- to 15-year planning horizon, as were areas having intact burrow systems identified in similar aerial photographs taken in 1983. The 1983 prairie dog colony distribution was used to estimate potential colony distribution because this was the period when recorded prairie dog populations were greatest. Other suitable prairie dog habitat areas lacking burrow systems since 1983 were not inventoried for this model under the assumption that population establishment in those areas was beyond the 10- to 15-year time frame of interest. Land areas within each survey section were classified as National Park Service–administered lands ($h = 1$), USDA Forest Service– administered lands initially subject to prairie dog population control (potential, $h = 2$), or USDA Forest Service–administered lands already reserved from prairie dog population control (current, $h = 3$). Privately owned lands were not included in the model.

Ferret Dispersal

Although few observations of ferret movements were available, distances of 2–3 km were typical for both nightly movements and annual intercolony movements (primarily by juveniles in late summer or early autumn) of wild-born ferrets at Meeteetse, Wyoming (Forrest et al. 1985; Biggins et al. 1986; Richardson et al. 1987). The longest nightly move reported from that complex was about 7 km. Oakleaf et al. (1992, 1993) report substantially longer dispersal distances (up to 17.5 km) over the first 30 days after captive-bred ferret releases at the Shirley Basin prairie dog colony complex in Wyoming. The statistics reported by Oakleaf et al. roughly suggest an exponential distribution of dispersal distances, and dispersal was apparently equally likely in all directions (although few observations were available). Eight of the ferrets released in Badlands National Park in 1994 were observed to disperse with a mean distance of 3.7 km and a maximum distance of 11.8 km (standard deviation = 4.2 km) over about a 30-day period. It was not known to what degree differences between these observations result from differences between captive-bred and wild-born ferrets, differences between prairie dog colony complexes, or other causes.

For this study, we assumed that all ferrets would disperse annually according to an exponential distance distribution with a mean of 3.7 km in uniformly random directions over a radius of about 14 km. We estimated dispersal coefficients (g_{ji}) by numerical approximation of the integral of this bivariate dispersal distribution over distances and angles defined by the boundaries of each destination (i) section relative to the center of each source (j) section. The effects of rugged topography in the Badlands could have been taken into account in the pairwise estimation of dispersal coefficients, but these effects were unknown.

Net Population Growth Rate

Wild ferrets have not been studied under conditions of unlimited habitat. Consequently, values for r_j were estimated by simulating unlimited population growth using mean birth and death rates and initial conditions similar to those assumed by Harris et al. (1989) in their research on black-footed ferret extinction probabilities. Beginning from expected values of 1 male (yearling or older), 1 yearling female, and 1.2 adult females (2 years or older), the simulated population was iteratively grown year by year. Yearling females were expected to produce .85 litters each, whereas adult females were expected to produce .95 litters each. Each litter was expected to produce 1.7 juvenile males and 1.7 juvenile females. Mortality then removed half the juvenile males, 40% of the juvenile females, 20% of the adult males, and 10% of the adult and yearling females. After 12 simulation years the population ratios and growth rates stabilized with an r value (annual net population growth rate) of .8175.

Ferret Releases

Based on past experiences (Oakleaf et al. 1992, 1993), approximately 80% of released captive-bred black-footed ferrets were expected to die during their first 30 days in the wild. Half of the remaining ferrets were expected to perish during their first winter. Of the 36 ferrets released in Badlands National Park in 1994, only 8 were known to survive the first 30 days. Taking into account likely winter mortality, we assigned an expected population value of .5 adult ferrets at each of the eight surviving ferret locations as initial conditions (N_i) in the model (with zeroes assigned elsewhere).

We expected 40 more ferrets to be released in the fall of each of the following 4 years. Assuming that four ferrets from each release would survive to reproduce, we set b_1 through b_4 equal to 4.0 (with zeroes assigned for all other years).

Ferret Carrying Capacity

Although prairie dog densities within colonies typically decline over extended periods of time (Cincotta 1985; Hoogland et al. 1988), initial colonies in the Badlands area were expected to remain well above the lower limit of good ferret habitat (3.63 prairie dogs/ha) estimated by Biggins et al. (1993) for the next 10–15 years. Consequently, we based our estimate of maximum ferret carrying capacity on the observations reported by Hillman et al. (1979) of ferret populations in Mellette County, South Dakota. In the model, adult ferret carrying capacities (c_{ihkt}) on existing or fully recovered prairie dog colonies were set at .05273 ferrets/ha. Figure 7.1 shows the spatial arrangement of initial plus potential ferret habitat by survey section in the study area at maximum model carrying capacity (determined by summing .05273 A_{ih} across h for each section i). Carrying capacity for the entire area was about 757 adult ferrets. Most of the ferret carrying capacity shown on the National Grassland (outside the National Park boundary) was potential rather than initial habitat, com-

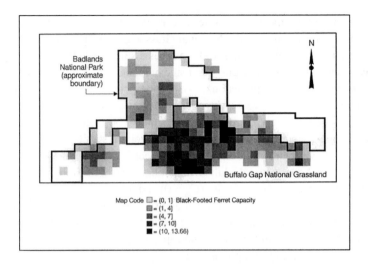

FIGURE 7.1

Adult black-footed ferret carrying capacities on federally managed lands
within the study area available in the short term (10–15 years) from 1994.

prising predominantly inactive prairie dog colony burrow systems. This did
not necessarily inhibit ferret establishment on the National Grassland in our
model, however, because prairie dog populations generally recover more
quickly than ferret populations can be established.

Based on studies by Knowles (1985b), Cincotta et al. (1987), and Apa et
al. (1990), we estimated that complete prairie dog population recovery in
recently treated colonies would take an average of four breeding seasons.
Therefore we set adult ferret carrying capacity (c_{ihkt}) at one-eighth of full
capacity (.00659 ferrets/ha) for the first year after use of rodenticide, one-
fourth of full capacity (.01318 ferrets/ha) for the second year, one-half of full
capacity (.02636 ferrets/ha) for the third year, and full capacity thereafter
(given no additional rodenticide treatments). This rate of recovery could
require special management actions to reduce vegetative cover, such as intensive livestock grazing, to aid the spread of prairie dogs (see Uresk et al. 1981;
Cincotta et al. 1988). We also assumed that all potential habitat areas in the
model could begin recovery in any chosen year.

Results

The model was solved with equation (7.9) as the objective function. Equation
(7.8) was used in six separate optimizations with different right-hand side

(C_{pt}) levels to limit the amount of ferret carrying capacity added from potential prairie dog colonies on the National Grassland ($h = 2$) to form a tradeoff
analysis. Thus for each of the six alternatives a single policy constraint ($p =$
1) was used with identical right- hand side amounts for each year t. The X_{i2k}
decision variables were given nonzero c_{ihktp} coefficients for years in which no
rodenticide treatments were scheduled. All other decision variables were
given c_{ihktp} coefficients of zero. A 25-year planning horizon ($T = 25$) was used
to allow enough time for ferret population levels to stabilize, but care was
taken not to overinterpret results beyond 15 years.

Figure 7.2 shows the total adult ferret population (F_t) resulting from allowing no additional carrying capacity ($C_{pt} = 0$) and from five 20% increments of
habitat (capacity for 103.88 additional adult ferrets per increment) from the
potential National Grassland prairie dog colonies. Because of ferret dispersal,
increments of additional ferret carrying capacity did not result in proportional
increases in expected ferret population.

In all cases in figure 7.2, sigmoid population growth curves resulted. As we
would expect, the graph shows diminishing marginal returns as more carrying capacity is added because the most spatially efficient habitat areas are
included first. Also, each curve levels off substantially below total allocated
carrying capacity (as in chapter 6). For example, when all National Grassland
habitat areas are allocated to prairie dog colonies, the expected population of
ferrets rises to only about 85% of the summed capacity of more than 757 adult
ferrets. This suggests that simply totaling available carrying capacity tends to
overestimate potential population sizes that can be supported because spatial
effects are not taken into account.

Preferred habitat areas change over time, as shown in figure 7.3 by maps
of the habitat allocated in different years (expressed as adult ferret capacity,
calculated by summing $c_{ihkt}X_{ihk}$ across h and k for each i and t) under the alternative that adds 20% of the potential National Grassland carrying capacity for
ferrets. The 20% limit for this alternative was binding from year 7 on. Before
that year, the expected ferret population was still small enough that the constraint was not limiting. Figure 7.3a shows the habitat allocations for year 7.
In figure 7.3a, a small amount of habitat was allocated to all but one survey
section with potential habitat because the fledgling ferret population was
rapidly expanding throughout the area (compare with figure 7.1). The
expected population in year 7 (S_{i7} for each section i) is shown in figure 7.4b,
along with the corresponding selected ferret releases (the sum of R_{it} across t
for each section i) shown in figure 7.4a.

By year 15, the expected ferret population under this alternative has largely
leveled off at more than 230 adult ferrets, and the preferred habitat allocations
have shifted to more concentrated areas around the National Park and initial

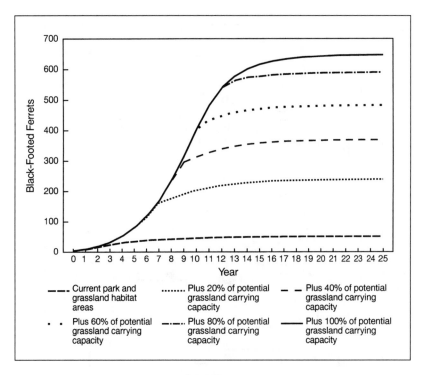

FIGURE 7.2

Projected optimal adult black-footed ferret populations under
the present management strategy and five alternative strategies.

National Grassland colonies (figure 7.3b). Many survey sections that had
some habitat allocated in year 7 no longer had any habitat allocated by year
15. The corresponding ferret population (S_{i15} for each section i) is shown in
figure 7.4c.

Based on these results, we examined the habitat allocations for this alternative near equilibrium conditions by bounding all ferret release variables to
zero (R_{it} in equation 7.5), unbounding all initial population variables (S_{i0} in
equation 7.3), and using equation (7.1) as the objective function. With initial
populations no longer restricting the solution, a static equilibrium could be
approximated without the need to build a different model. We also allowed
the model to schedule treatments for National Grassland colonies initially
reserved from rodenticide use ($h = 3$) in addition to potential colony ($h = 2$)
treatments for meeting the policy constraint to allow greater spatial freedom
for habitat selection. The resulting allocations (figure 7.3c) support a popula-

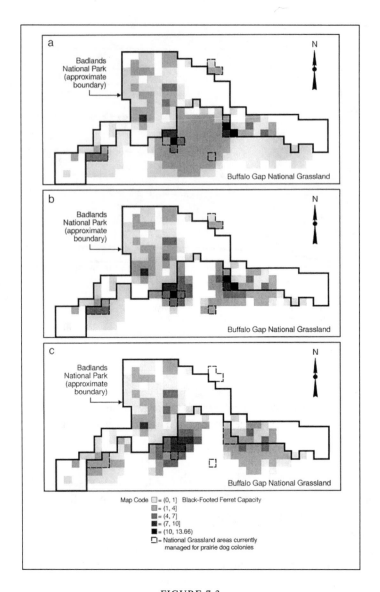

FIGURE 7.3

Habitat allocations expressed as adult black-footed ferret carrying capacities under the +20% alternative (a) in year 7, (b) in year 15, and (c) near equilibrium.

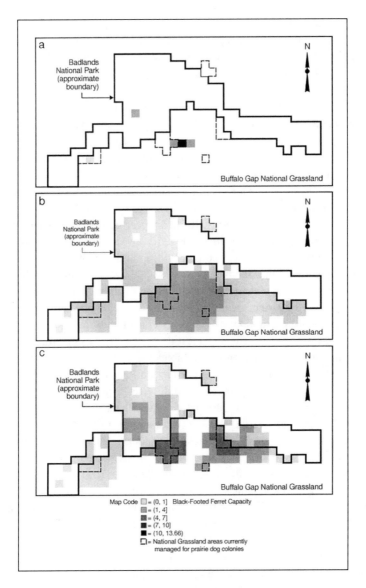

FIGURE 7.4

The number of adult black-footed ferrets by location under the +20%
alternative (a) released at selected sites, (b) in year 7, and (c) in year 15.

tion that levels off at a little more than 300 adult ferrets. This is an increase of about 60 adult ferrets (in year 25) over the solution in figure 7.2. The increase is achieved by exchanging some (but not all) of the prairie dog colonies from the initial reserve areas for new colonies in the nearby National Grassland areas (compare figures 7.3b and 7.3c).

The pattern of National Grassland allocations in figure 7.3c results from the interaction of two important effects. First, many of the sections with the highest potential ferret capacity were left unallocated by the model to round out the long, narrow habitat arrangements in portions of the National Park. Second, this tendency appears to be compromised somewhat in favor of placing habitat close to as many sections of the National Park (where habitat is fixed) as possible.

Public land use planning for a 10- to 15-year period often is viewed as a process for scheduling a one-time change (if any) in management. To test the effects of that approach, we constructed a model in which the potential National Grassland management variables (X_{i2k}) were redefined such that the model could choose when to stop rodenticide treatments in a particular area (if at all), and no further treatments could be scheduled afterward. In some planning cases, scheduling is not even considered, and all management changes take immediate effect. To test those effects, we constructed another model in which the management variables for each potential section (again, X_{i2k}) simply represented either repeated rodenticide treatments or none at all. Table 7.1 shows the yearly expected adult ferret populations (F_t) using these different approaches for the alternative that allocates 20% of the potential National Grassland carrying capacity (in addition to the National Park and current National Grassland colonies). Considering the small differences in table 7.1, the use of simpler models (with greater ease of presentation and implementation) may not affect results significantly. The reductions in matrix size and complexity with the simpler models were substantial. The full scheduling formulation included 20,946 rows and 65,598 columns, whereas the simpler "one-time change in management" scheduling model reduced the problem size to 17,795 rows and 21,972 columns. The nonscheduling model further reduced the number of columns to 18,156.

With very limited knowledge of ferret reproduction and dispersal in the wild, our model results were regarded as an initial estimate of a lower bound on expected population levels for a given habitat arrangement. Stochastic variation, which can be a particularly important consideration at low population levels, was not taken into account. The explicit accounting of spatial patch relationships, as opposed to relying on measures such as mean intercolony distance, is the strong point of the model. Viewing the model's population esti-

TABLE 7.1

Expected Number of Black-Footed Ferrets in Each Year
for the +20% Alternative Under Three Different Scheduling Formulations

Year	Allocation in Year 1	One-Time Allocation Change in Any Year	Full Scheduling Model
1	9	9	9
2	17	17	18
3	31	31	32
4	51	51	53
5	76	76	79
6	106	107	114
7	139	139	161
8	163	169	177
9	182	183	190
10	196	195	202
11	205	204	212
12	212	210	219
13	217	216	225
14	220	219	229
15	222	222	232
16	224	224	234
17	225	225	236
18	225	226	237
19	226	227	238
20	226	227	238
21	227	228	239
22	227	228	239
23	227	228	239
24	227	228	240
25	227	228	240

mates as lower bounds provided a useful contrast to results from habitat complex circumscription methods (e.g., Biggins et al. 1993), which could probably be viewed as estimates of expected population upper bounds, at least without qualitative adjustments. In chapter 8 we examine whether the prairie dog colony arrangements proposed here for ferret recovery are suitable for the prairie dogs themselves.

8

POPULATION-DEPENDENT DISPERSAL: HABITAT PLACEMENT FOR THE BLACK-TAILED PRAIRIE DOG

The Black-Tailed Prairie Dog

In the late 1800s, an estimated 283 million ha were occupied by a combined population of North American prairie dogs (*Cynomys* spp.) exceeding some 5 billion individuals (Seton 1929). By 1971, that area had declined to 600,000 ha (Fagerstone and Biggins 1986). Loss of habitat, control programs, and plague (*Yersinia pestis*) continued to reduce populations to the point that it is currently estimated that prairie dogs have been reduced by 98–99% of their former numbers across the western United States (Miller et al. 1994; Mulhern and Knowles 1997). Ever since Merriam (1902) reported that prairie dogs compete with livestock for forage, they have been targeted for control programs as agricultural pests. More recently, however, prairie dogs have been viewed as keystone species important to many other species that depend on them for food, burrows, and their effects on plant communities (Miller et al. 1994). Nonetheless, rodenticide control programs probably will continue because of the competition between prairie dogs and livestock (Roemer and Forrest 1996; Long 1998).

The black-tailed prairie dog (*C. ludovicianus*) was historically the widest-ranging of the five prairie dog species in North America, occupying areas from Mexico to Canada. Members of this species are the only prairie dogs found on the Great Plains (Wuerthner 1997). South Dakota is a particularly impor-

---

This chapter was adapted from J. Hof, M. Bevers, D. W. Uresk, and G. L. Schenbeck, Optimizing habitat location for black-tailed prairie dogs in southwestern South Dakota, *Ecological Modelling* 147 (2001): 11–21, with permission from the publisher, Elsevier Science.

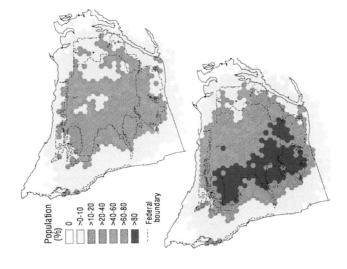

PLATE 1

Current owl habitat available for retention (top)
and habitat in the plan to be evaluated (bottom).

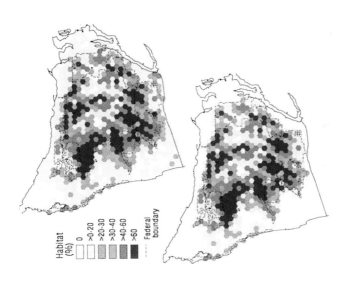

PLATE 2

Simulation of mean owl population (mean occupancy
of each hexagonal cell) with all current habitat,
with rule set b (top) and rule set d (bottom).

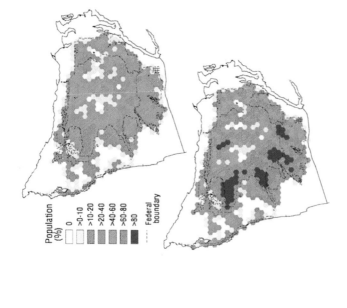

PLATE 3

Mean owl population (mean occupancy of each hexagonal cell) from the optimization model with all current habitat, with rule set b (top) and rule set d (bottom).

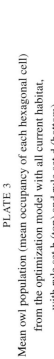

PLATE 4

Mean owl population (mean occupancy of each hexagonal cell) from the optimization model with the habitat retained in the plan, with rule set b (top) and rule set d (bottom).

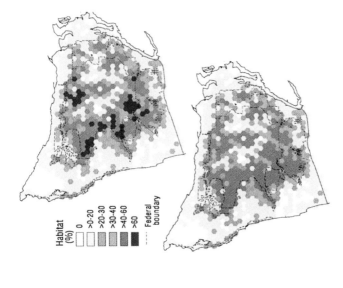

PLATE 5

Mean owl population (mean occupancy of each hexagonal cell) from the simulation model with the habitat retained in the plan, with rule set b (top) and rule set d (bottom).

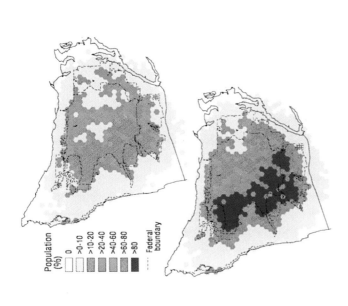

PLATE 6

Optimized owl habitat with total habitat constrained to the amount in the plan, with rule set b (top) and rule set d (bottom).

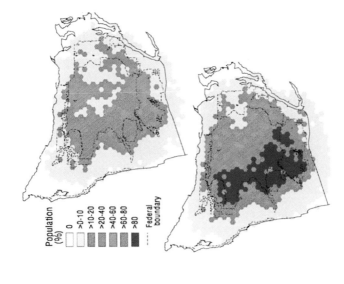

PLATE 7

Optimized mean owl population (mean occupancy of each hexagonal cell) with total habitat constrained to the amount in the plan, with rule set b (top) and rule set d (bottom).

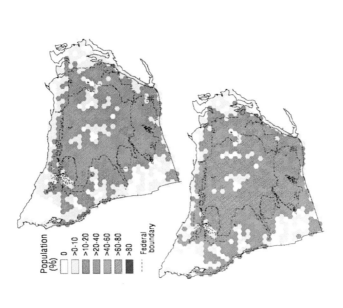

PLATE 8

Mean owl population (mean occupancy of each hexagonal cell) from the simulation model, with the optimized habitat layouts from plate 6 imposed, with rule set b (top) and rule set d (bottom).

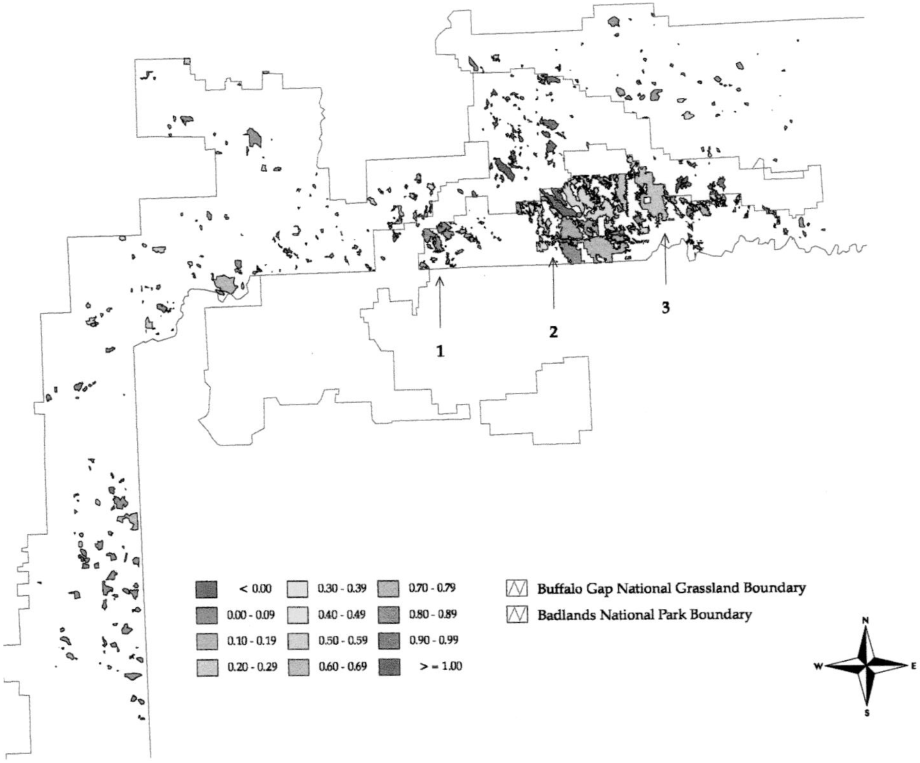
1
2
3
< 0.00
0.00 - 0.09
0.10 - 0.19
0.20 - 0.29
0.30 - 0.39
0.40 - 0.49
0.50 - 0.59
0.60 - 0.69
0.70 - 0.79
0.80 - 0.89
0.90 - 0.99
> = 1.00
Buffalo Gap National Grassland Boundary
Badlands National Park Boundary
N
W
E
S

PLATE 9
Map of the ratio of currently occupied prairie dog habitat area to total prairie dog habitat area in the Buffalo Gap National Grassland and the Badlands National Park.

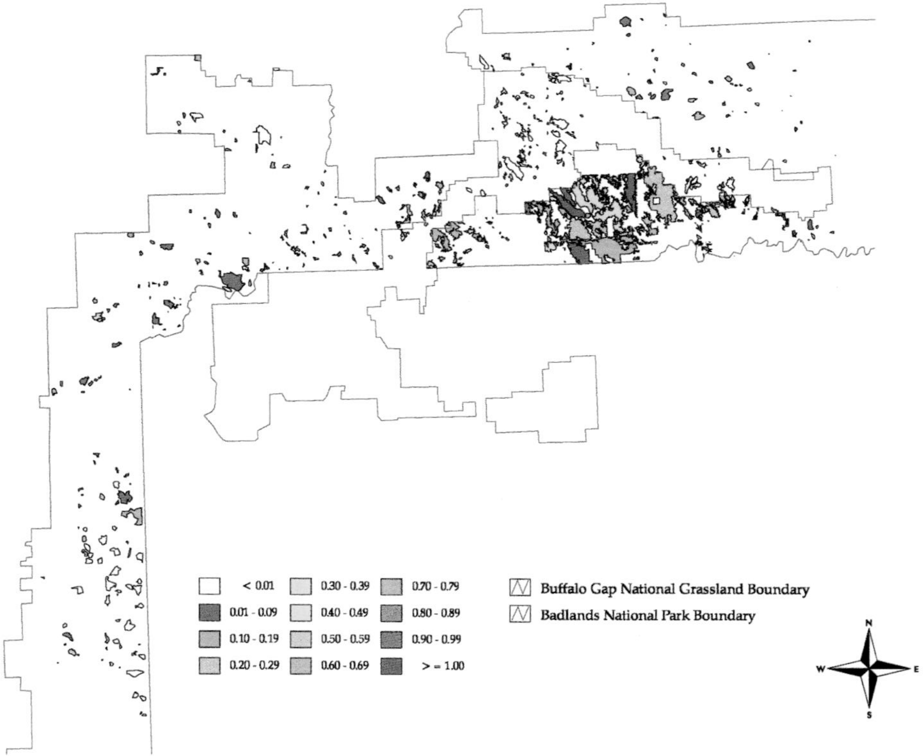

PLATE 10
Additional protected habitat area
÷ total habitat area in the +25%
optimization solution.

Buffalo Gap National Grassland Boundary
Badlands National Park Boundary

< 0.01	0.30 - 0.39	0.70 - 0.79
0.01 - 0.09	0.40 - 0.49	0.80 - 0.89
0.10 - 0.19	0.50 - 0.59	0.90 - 0.99
0.20 - 0.29	0.60 - 0.69	>= 1.00

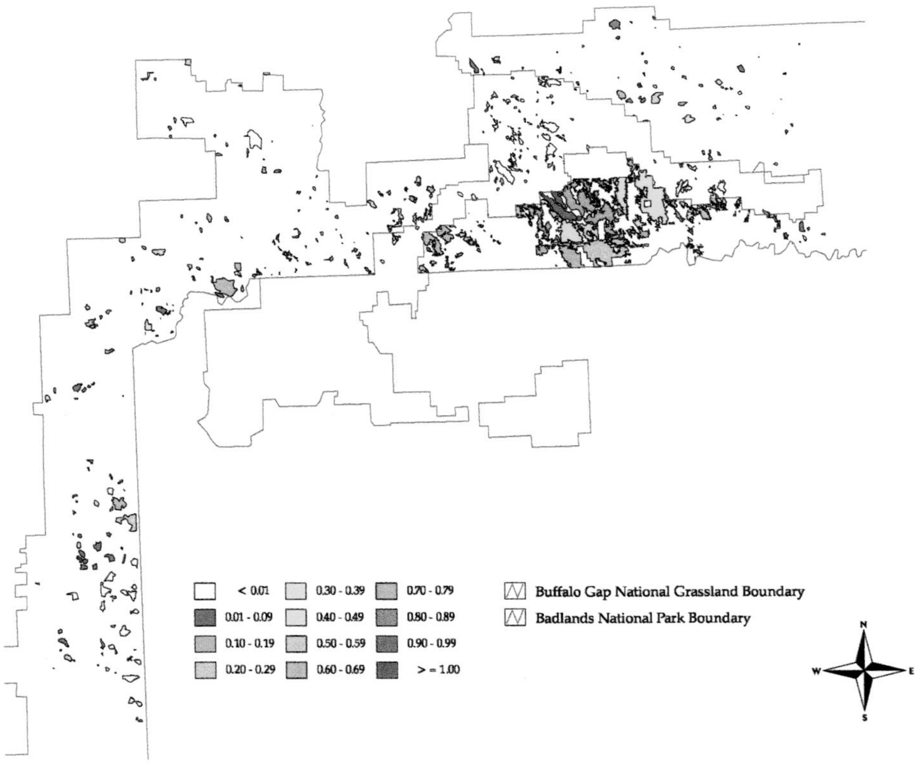

PLATE 11

Additional protected habitat area ÷ total habitat area in the +50% optimization solution.

Buffalo Gap National Grassland Boundary

Badlands National Park Boundary

< 0.01	0.30 - 0.39	0.70 - 0.79
0.01 - 0.09	0.40 - 0.49	0.80 - 0.89
0.10 - 0.19	0.50 - 0.59	0.90 - 0.99
0.20 - 0.29	0.60 - 0.69	> = 1.00

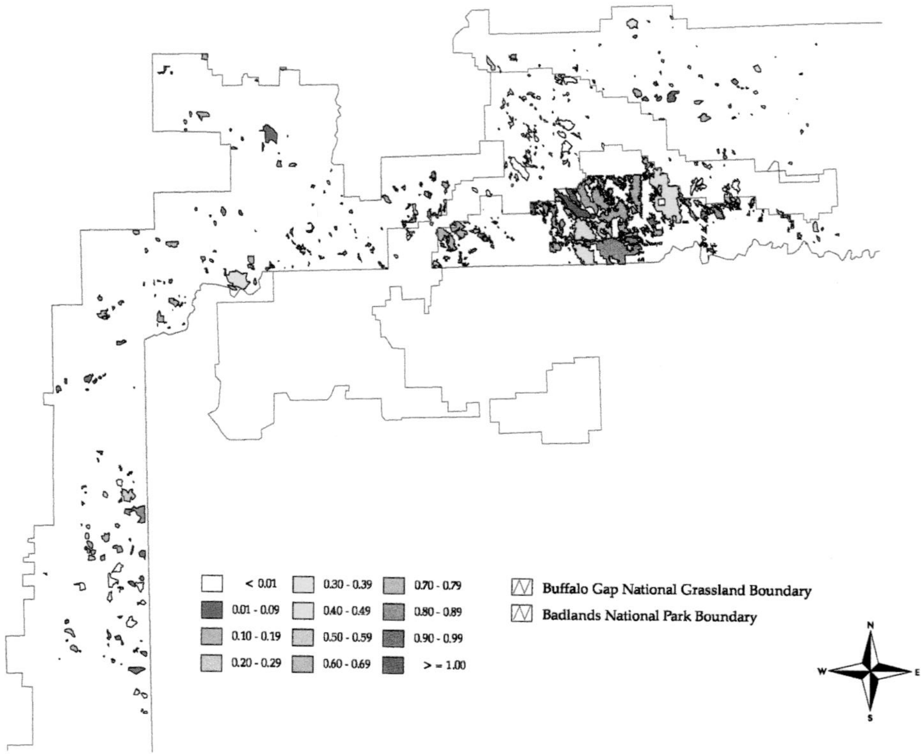

PLATE 12
Additional protected habitat area
÷ total habitat area in the +75%
optimization solution.

Buffalo Gap National Grassland Boundary

Badlands National Park Boundary

< 0.01	0.30 - 0.39	0.70 - 0.79
0.01 - 0.09	0.40 - 0.49	0.80 - 0.89
0.10 - 0.19	0.50 - 0.59	0.90 - 0.99
0.20 - 0.29	0.60 - 0.69	> = 1.00

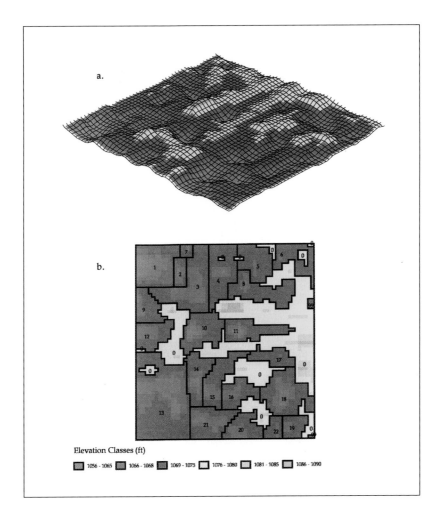

a.

b.

Elevation Classes (ft)

1056 - 1065   1066 - 1068   1069 - 1075   1076 - 1080   1081 - 1085   1086 - 1090

PLATE 13
Landscape of the case study area (a)
and the delineation of numbered swale complexes (b).

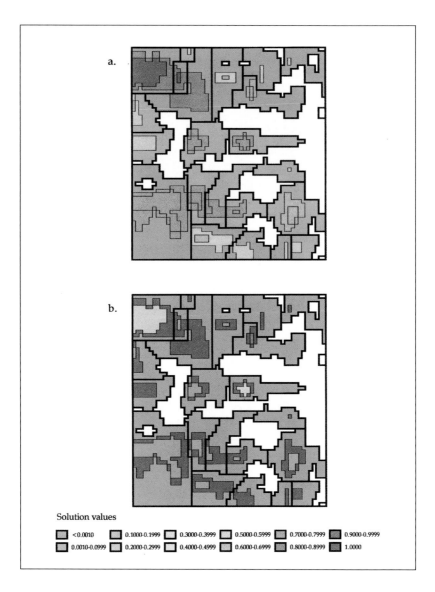

a.

b.

Solution values

| | <0.0010 | | 0.1000-0.1999 | | 0.3000-0.3999 | | 0.5000-0.5999 | | 0.7000-0.7999 | | 0.9000-0.9999 |
| | 0.0010-0.0999 | | 0.2000-0.2999 | | 0.4000-0.4999 | | 0.6000-0.6999 | | 0.8000-0.8999 | | 1.0000 |

PLATE 14

Area allocated to orchid habitat (white areas are nonhabitat) in the base solution
with initial seed dispersal assumptions in time period 16 (a) and time period 20 (b).

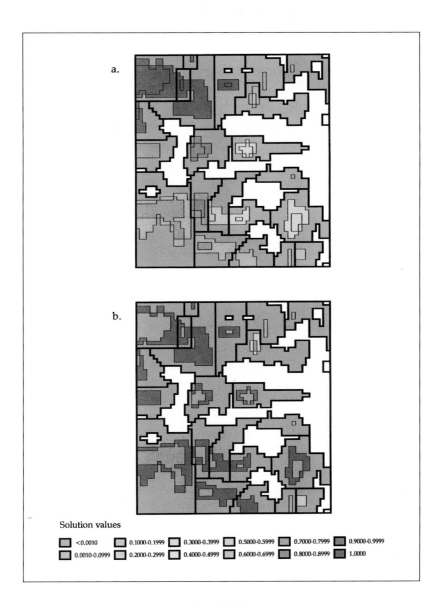

PLATE 15
Area allocated to orchid habitat (white areas are nonhabitat)
in base solution with conservative seed dispersal assumptions
in time period 20 (a) and time period 25 (b).

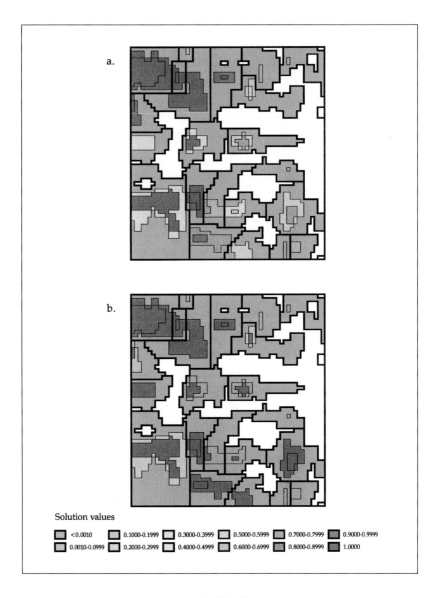

Solution values

■ <0.0010	■ 0.1000-0.1999	■ 0.3000-0.3999	■ 0.5000-0.5999	■ 0.7000-0.7999	■ 0.9000-0.9999
■ 0.0010-0.0999	■ 0.2000-0.2999	■ 0.4000-0.4999	■ 0.6000-0.6999	■ 0.8000-0.8999	■ 1.0000

PLATE 16
Area allocated to orchid habitat (white areas are nonhabitat) in solution
with altered climate probabilities and with conservative seed dispersal assumptions
in time period 20 (a) and time period 25 (b).

tant location for black-tailed prairie dogs because of the apparent absence of plague epizootics (Barnes 1993; Fitzgerald 1993). Thus public interest in efforts to increase the prairie dog numbers on public lands in South Dakota is high (Long 1998). Previously occupied prairie dog habitat often is repopulated if protected from rodenticide treatments, but livestock grazing may also be necessary to aid in the spread of prairie dog populations by reducing vegetative cover in selected areas (Uresk et al. 1981; Cincotta et al. 1988).

Given that land areas for additional prairie dog colonies probably will be limited under any conservation strategy, an important question is, "Where should any additional colonies be located?" The spatial arrangement of protected habitat has important effects on the ability of the prairie dogs to occupy and use that habitat (Hoogland et al. 1995; Garrett and Franklin 1982). This chapter describes the use of optimization methods to find efficient spatial arrangements of protected habitat for expanding prairie dog populations on the Buffalo Gap National Grassland.

In chapter 7, we modeled the releases and population recovery of an associated species, the black-footed ferret (*Mustela nigripes*). In that model, prairie dog colony placements were treated as choice variables, and the ferret population response reflected its total dependence on prairie dogs for food and shelter. We characterized the spatial options with a grid of uniform cells and captured the dispersal behavior of the aggressively dispersing ferret with a set of reaction–diffusion relationships. The prairie dog is much more hesitant to disperse and typically does so only if population densities approach habitat patch carrying capacity (Cincotta et al. 1988). Therefore to model the prairie dog we characterize the landscape choice variables with irregular patches of habitat and the dispersal behavior with a population-dependent yet linear reaction–diffusion formulation. The two models therefore are very different in terms of structure and formulation. Results from both models are compared to investigate the hypothesis that managing for prairie dogs (given their population dynamics) is (or is not) consistent with managing for ferrets (given their population dynamics). We begin by discussing prairie dog habitat and the population dynamics we wanted to capture in the optimization model.

The Model

As in chapter 7, we identified patches of black-tailed prairie dog habitat from maps of prairie dog presence during the times of highest known populations. Those patches were assumed to be the only part of the landscape that could support prairie dog populations within the planning time horizon (8 years), so they limited our choice variables. The choice variables determine the area within each patch of habitat that is zoned and protected for prairie dog colonies. It was assumed that grazing regimes conducive to prairie dog presence would be applied in the areas zoned for colonies, and poisoning would be eliminated. Areas in the patches of habitat that were not zoned for colonies were assumed to be subject to poisoning and other management actions that make them unsuitable for prairie dog occupation.

To model prairie dog populations, we had to make some specific assumptions about their dynamics. First, we assumed that at any time, prairie dog populations in any colony are determined by whichever is limiting: population growth and dispersal processes or the protected habitat carrying capacity determined by our choice variables. Second, we assumed that if unconstrained by rodenticides, prairie dog populations grow exponentially up to habitat capacity with an r value that accounts for resident natality and mortality.

Third, the only prairie dog movements we modeled were those that resulted in net dispersal between habitat patches. This means that we ignored intra-patch movements and animal exchanges between patches for breeding purposes (which roughly balance out). Hoogland (1995:87–88) states, "Emigration to, and colonization of, new colony sites probably is expensive and dangerous for prairie dogs. . . . However, expansion of already existing colonies should be easier and safer." Also, Cincotta et al. (1988:31) state, "Results indicated that prairie directly adjacent to the study colony was likely to be colonized if it was near a dense population of prairie dogs. . . . High population densities may force prairie dogs to expand into new territory." Thus prairie dogs appear to be reluctant to disperse from patch to patch. Prairie dogs probably begin to disperse to other areas before their colonies become completely saturated, but as a conservative modeling approach we assumed that no dispersal emigration occurs until habitat areas are at carrying capacity, determined with a fixed parameter (per area of habitat).

Fourth, we assumed that once dispersal occurs, it is random. Therefore we identified a probability that a dispersing prairie dog from each patch will successfully disperse to any other patch. This probability could reflect barriers to dispersal (and other landscape features) but is simply a function of the distance between the two patches in this exploratory chapter (see Garrett and Franklin 1982). Our model is structured so that dispersal between all patches of habitat is determined simultaneously for each 1-year time period. Implicit in this simultaneous solution is the feature that if prairie dogs attempt to enter a habitat area that is already full, those animals return to the pool of dispersers and are again subjected to the dispersal mortality associated with moving to other patches.

Fifth, we assumed that prairie dogs reproduce each year (and the r value is applied) before they disperse. The literature also suggests that dispersing prairie dogs do not reproduce during their first year in a new colony. As Cincotta et al. (1987:341) state, "There was no observed reproductive success during either 1982 or 1983 among the newly established populations of prairie dogs on the treated site. This suggests that female black-tailed prairie dogs disperse after the mating season and do not bear young during their first year in a newly established territory." This temporary nonmating behavior is a subtlety that we did not capture in our linear optimization model. Instead, we assumed that all prairie dogs reproduce similarly. The combination of highly population-dependent dispersal and random dispersal once it does occur still created a conservative model of prairie dog colonization.

With these assumptions, our model is formulated as follows:

Maximize

$$\sum_{t=1}^{T}\sum_{i=1}^{M} S_{it}, \tag{8.1}$$

subject to

$$X_i \leq A_i \qquad \forall i, \tag{8.2}$$
$$S_{i1} = N_i \qquad \forall i, \tag{8.3}$$
$$Q_{i1} = 0 \qquad \forall i, \tag{8.4}$$
$$S_{it} + Q_{it} \leq (1+r)S_{i(t-1)} + \sum_{j\neq i} g_{ij}Q_{jt} \quad \begin{matrix} \forall i \\ t = 2,\ldots,T, \end{matrix} \tag{8.5}$$
$$S_{it} \leq C_i X_i \qquad \begin{matrix} \forall i \\ \forall t, \end{matrix} \tag{8.6}$$

$$\sum_{i=1}^{M} X_i \leq B. \tag{8.7}$$

Indexes

i indexes patches, as does j,
M = the number of patches,
t indexes time periods,
T = the number of time periods.

Variables

S_{it} = the prairie dog population in patch i, time period t,
X_i = the number of hectares in patch i zoned for prairie dog colonies,

Q_{it} = a variable used to account for the prairie dogs that leave patch i in time period t to disperse to other patches. The number of prairie dogs that successfully disperse to patch i from other patches ($j \neq i$) is

$$\sum_{j \neq 1} g_{ij} Q_{jt}$$

(see text for explanation).

Parameters

A_i = the number of hectares of prairie dog habitat in patch i,
N_i = the initial population of prairie dogs in patch i,
r = the r value reproduction rate for resident prairie dogs,
g_{ij} = the probability of a dispersing prairie dog leaving patch j and successfully reaching patch i,
C_i = the carrying capacity of prairie dogs per hectare of protected habitat,
B = a limit on the total amount of area that can be zoned for prairie dog colonies (a policy parameter).

Equation (8.1) is the objective function and maximizes the total prairie dog population over all patches and time periods. Equation (8.2) limits the zoned colony area to the habitat area in each patch. Equation (8.3) sets the population in the first time period to the initial population. Equation (8.4) initializes the number of dispersing prairie dogs (Q_{i1}) at zero. The population in each patch and time period, S_{it}, is determined by whichever of constraints (8.5) or (8.6) (or (8.3) for the first time period) is binding. Equation set (8.5) determines the population of each patch i in each time period t, if it is binding, as follows: The population from the previous time period ($t - 1$) for patch i, $S_{i(t-1)}$, is expanded by the reproduction factor $(1 + r)$. Then, the prairie dogs successfully immigrating from other patches ($j \neq i$),

$$\sum_{j \neq i} g_{ij} Q_{jt},$$

are added. Equation (8.6) limits the prairie dog population to the habitat carrying capacity determined by the solution values for X_i. If equation (8.6) is binding for a given i, then the excess population in (8.5) is accounted for in the Q_{it} variable, which simultaneously enters into the other equations in (8.5) as dispersing prairie dogs. The model solution always sets $Q_{it} = 0$ until equation (8.6) is binding because the sum of the S_{it} is being maximized and

$$\sum_{i \neq j} g_{ij}$$

for any patch j is always less than 1, reflecting dispersal-related mortality. Thus no dispersal occurs until allocated habitat (X_j) is full, as desired. Equation set (8.5) is a simultaneous system of equations that will solve for all Q_{it} and Q_{jt}, accounting for all dispersal between all patches in the given time period. Equation (8.7) limits the total prairie dog colony area across all patches as a policy constraint.

Implicitly, prairie dog emigration is modeled as a multistep process with mortality occurring (accounted for in the g_{ij} coefficient) at each step. The more steps it takes a given prairie dog to find an unoccupied patch, the more mortality risk accumulates (the g_{ij} coefficients apply at each step). We defined a maximum dispersal distance of 5 km (Garrett and Franklin 1982, as cited by Hoogland 1995) for each step. Thus to define the g_{ij} coefficients we drew a 5-km radius around the centroid of each patch j. We assumed that any patch i whose centroid is within that radius had an equal chance of being located by a prairie dog emigrating from patch j. The actual dispersal process may be affected by a number of factors including corridors of high visibility, which the prairie dogs prefer; the chirping of other prairie dogs, which may serve as an attractant; and other biological and landscape variables. These processes are not well understood, and observed prairie dog dispersal behavior is rare (Hoogland 1995). Thus random dispersal direction seemed to be the most tenable assumption. We applied a mortality rate of 40% more than that already accounted for in the r value (Garrett and Franklin 1982) to calculate all g_{ij}. In effect then, modeled prairie dogs can disperse farther than 5 km (consistent with Knowles 1985a and others) given a suitable network of patches but with an approximately exponential decay in survival of 40% for each 5 km. The dispersal distance and mortality rate both come from the same study (Garrett and Franklin 1982), which took place in an area of rapid expansion and plentiful habitat. We thus assumed that these figures were applicable to each step because in the Garrett and Franklin study only the first step was likely to be observed (few emigrating prairie dogs encountered fully populated patches).

We adopted an r value of .4 (so $1 + r = 1.4$), which is at the lower end of the range reported by Hoogland (1995). Given the presence of hunting and many predators (including the black-footed ferret), this seemed appropriate. A carrying capacity (C_i) of 18.4 prairie dogs per hectare was assumed for all habitat patches and all time periods. This was the average observed population density for mature colonies measured (with repeated trapping) over 2 years on 12 plots in the study area (personal communication, Kieth E. Severson).

Plate 9 shows the study area with 601 habitat patches covering approximately 20,443 ha. Initial occupancy (unpublished data, Nebraska National

Forest, updated summers of 1996 and 1997) covered approximately 7991 ha and is indicated with color coding. As noted earlier, the habitat was determined from historically high populations. The patches in Badlands National Park were assumed to be at capacity initially. Thus the only choice variables for additional protected habitat were in the Buffalo Gap National Grassland, but the model includes the Badlands National Park patches to account for dispersal from that population into other areas. On the National Grassland, prairie dog densities were initially highest in three areas: the red patches labeled 1 in plate 9, the red and orange patches labeled 2 in plate 9, and the orange patch labeled 3 in plate 9. These areas had been previously designated as protected prairie dog habitat. Poisoning had been allowed to varying degrees elsewhere in the National Grassland, with the population varying in the initial conditions as indicated. Areas 2 and 3 are at the western and eastern limits of an area called Conata Basin. The model was built with eight 1-year time periods (with one exception discussed later). With these dimensions, the model had 9626 choice variables and 8424 constraints.

Results

Four policy alternatives were examined by solving the model with different right-hand side (B) levels in constraint (8.7) that represent protecting 25%, 50%, 75%, and 100% of the initially unoccupied habitat in addition to what was already occupied. Each 25% increment amounts to an increase of 3113 ha, which supports 57,280 prairie dogs at a carrying capacity of 18.4 per ha. The population results are given in figure 8.1. With the first three levels of increased prairie dog habitat, all available habitat was occupied within 8 years. With 100% of the habitat protected, all but about 4.5 ha eventually were occupied, but a 21-year model was needed before no further population increases occurred. These 4.5 ha initially were unpopulated and were isolated by the geography and dispersal assumptions in the model. Essentially all the habitat can be fully occupied with enough time. The optimization model therefore is most useful for identifying the patches of additional protected habitat that can be populated by prairie dogs most quickly for a given policy alternative.

Table 8.1 compares the optimized rates of population increase with the population increases from the model solved with "neutral" protected locations imposed. For the neutral solutions, the given percentage (25%, 50%, or 75%) of the habitat that is initially unpopulated is protected in all patches. This represents a neutral spatial allocation in that an equal proportion of additional habitat in all patches is protected, analogous to the random selections of null models in ecology (Gotelli and Graves 1996). As can be seen in table 8.1, the

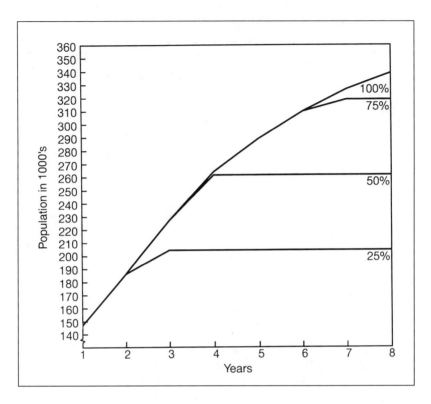

FIGURE 8.1

Prairie dog population trends for four levels of habitat increase
(+25%, +50%, +75%, and +100% of currently unpopulated habitat).

optimized location of habitat additions supports more rapid prairie dog population increases (at least as indicated by the model itself).

The optimized protected habitat locations chosen by the model for the three
habitat expansion levels in table 8.1 are presented in plates 10–12. Plate 10
shows the habitat additions selected for each patch as proportions of total
patch sizes for the +25% solution. The highest priority area for habitat expansion was indicated to be in Conata Basin. The yellow area surrounding area 2
in Conata Basin was the area with the largest habitat increases in solution
(compare with plate 9). Thus the top priority indicated was to expand the size
of area 2 to the east and south. This top priority area does not go all the way
south to the National Grassland boundary, so a buffer between large colonies
and other land ownerships probably could be maintained without much loss.

TABLE 8.1

Model Prairie Dog Populations (in 1000s) Over Time
for Spatially Optimized and Spatially Neutral Patch Allocations
Under Three Increments of Habitat Increase

Habitat Increase	+25%		+50%		+75%	
Year	Optimum	Neutral	Optimum	Neutral	Optimum	Neutral
1	147.0	147.0	147.0	147.0	147.0	147.0
2	186.1	175.2	186.1	182.4	186.1	185.0
3	204.3	186.1	227.1	208.0	227.1	220.2
4	204.3	192.5	261.6	222.9	264.1	245.8
5	204.3	195.7	261.6	234.6	289.3	263.9
6	204.3	197.5	261.6	241.9	309.9	279.0
7	204.3	198.7	261.6	245.7	318.9	289.3
8	204.3	199.7	261.6	248.5	318.9	294.7

The next priority was indicated to be increasing the level of protection in areas 1 and 3 (note that it was impossible to expand the size of area 1 very much). Also, it was clearly desirable to connect areas 2 and 3 more than they were initially. It is actually surprising that even with a recalcitrant disperser such as this animal, it was still optimal to expand the habitat significantly outside Conata Basin. In particular, the area north and east of the National Park, the area due west of Conata Basin, the area due east of area 3, and the middle part of the southwestern leg of the study area all were increased in the optimal solution. The areas in the extreme southwest and northwest of the grassland were not indicated to be high priority. This solution thus suggested a strategy in which we try to protect some additional habitat in the southwest National Grassland, northwest National Grassland, and the National Grassland northeast of the National Park; we give high priority to expanding area 2 into a nearly round, coterminous area in Conata Basin that will serve as the core area for the population; and we also try to connect area 1 enough to make it a part of this core area. Area 3 probably is also very important for connecting this core area to the west and southwest fringe areas. The National Park serves this purpose for the northeast Grassland. As one might expect, the population trends and locations followed the patterns of habitat location, with all available habitat fully occupied by year 3 (table 8.1).

Plate 11 shows the habitat added to each patch as a proportion of total patch size for the +50% solution. This solution is almost the same as the +25% solution west of the core area around area 2, although there is some additional protected habitat in the southwest leg of the study area. The emphasis in this solution is on increasing the density in the core area and expanding it to the point that it encompasses area 3 and the patches east of there. Some priority is also given to the northeast part of the study area.

Plate 12 shows the habitat added to each patch as a proportion of total patch size for the +75% solution. This solution expands the core area south to the National Grassland boundary. The core area is now expanded to the limits of the habitat locations. Area 1 still is not given high priority, and the northwest areas are still at very low densities. Some high densities are now protected in the far southwest, however, as well as the patches due west of the core area. The patches in the northeast part of the National Grassland now have protected densities as high as 100%, especially close to the National Park boundary and the initial high densities found there.

In summary, all three solutions emphasize the core area but not completely at the expense of expanding the overall size of the total population area. Our model indicated that optimally located habitat additions could be used more rapidly to expand prairie dog populations than a neutral locational scheme (table 8.1). The initial areas protected from rodenticide use were well located for this potential, especially areas 2 and 3.

Our model required many assumptions regarding the population dynamics of the black-tailed prairie dog and did not explicitly account for random variation in those dynamics. Of particular importance is the possibility of a plague epizootic or other catastrophic event that could invalidate our model results. Therefore our results were regarded as an initial estimate of the expected population levels for the optimized habitat arrangements. Although the indicated improvements in total population from using a spatial optimization model are modest (table 8.1), the arrangement strategies suggested by our results appear to be useful in hastening the use of habitat additions by prairie dogs.

We have contrasted the formulations in this chapter with those in chapter 7 that modeled the black-footed ferret. The management strategies emerging from the two models are quite consistent, however. Our ferret model used dynamic (scheduled) choice variables as opposed to the fixed zoning choice variables in the prairie dog model, and the ferret model covered only the northern part of the area modeled for the prairie dogs. Still, both models indicate that initially protected prairie dog colony areas are well selected but should be expanded if increased populations are desired. The strategy from the ferret model was to add a little habitat everywhere at first and then later to concentrate the habitat in Conata Basin. Though fixed, the strategy suggested here is similar: Create a core area in Conata Basin but also increase other areas to expand the geographic extent of the population. Thus it appears that locating prairie dogs to maximize ferret populations taking ferret population dynamics into account does create a similar strategy to maximizing prairie dog populations taking prairie dog population dynamics into account. Put another way, the optimized strategy in managing for ferret recovery is plausible given prairie dog dynamics, and managing for the prairie dogs themselves is not inconsistent with our objectives of supporting ferrets. In chapter 9, we model a prairie plant species that exhibits multiple life stages and a dispersal process that is strongly influenced by the combination of topography and precipitation.

9

TOPOGRAPHY-BASED DISPERSAL: HABITAT LOCATION FOR THE WESTERN PRAIRIE FRINGED ORCHID

Spatial patchiness is a common feature of the distributions of most plants, particularly those occurring in ephemeral habitats (e.g., Schemske et al. 1994). This factor combined with their immobility through certain stages of life makes plants seem particularly appropriate for spatial landscape analyses. A surprisingly small number of studies have examined plant populations with such an approach, however, particularly in terms of testing theoretical models (Husband and Barrett 1996; Wu and Levin 1997). Plant population dynamics differ from those of animals in a number of ways, including vulnerability levels independent of the size and age of populations (e.g., Husband and Barrett 1996) and the role of seed dormancy in maintaining populations (e.g., Kalisz and McPeek 1993). A potential outcome of these factors, especially in ephemeral habitats such as wetlands, is that local changes in habitat conditions can cause drastic reductions in population size and distribution; but the presence of a viable seed bank can save the species from extinction. Therefore one of the challenges in developing conservation strategies for rare plants is to develop plans that account for the influence of climatic conditions and seed viability and dispersal through time on colonization and extinction rates of local populations (Malanson and Armstrong 1996; Malanson 1996; Valverde and Silvertown 1997).

This chapter was adapted from J. Hof, C. H. Sieg, and M. Bevers, Spatial and temporal optimization in habitat placement for a threatened plant: The case of the western prairie fringed orchid, *Ecological Modelling* 115 (1999): 61–75, with permission from the publisher, Elsevier Science.

This chapter explores the possibility of using an optimization modeling approach to maximize a plant species population by strategically selecting the placement and timing of protected plant habitats, using a threatened wetland species as an example. Optimization shows promise for this problem because efficiency is important, given the limited amount of habitat that can be protected (Nevo and Garcia 1996; Batabyal 1998; Tiwari et al. 1996).

The Western Prairie Fringed Orchid

The western prairie fringed orchid (*Platanthera praeclara* Sheviak and Bowles) was once distributed throughout the wetlands west of the Mississippi River in the tallgrass prairie of the central United States and southern Canada. With the conversion of prairie to cropland and the encroachment of other human development, much of the original habitat has been lost. Concern about the limited number of populations and their relative isolation led to federal listing of the species as threatened (U.S. Fish and Wildlife Service 1989). Populations of *P. praeclara,* referred to as "the orchid" hereafter, are limited to the Great Plains States and Manitoba; the three largest metapopulations occur in Minnesota and North Dakota in the United States and in southern Manitoba in Canada (U.S. Fish and Wildlife Service 1996).

The orchid metapopulation in North Dakota is centered on the Sheyenne National Grassland and nearby roadside ditches and private lands in the southeastern corner of the state (U.S. Fish and Wildlife Service 1996). Within this area, the orchid has a patchy distribution that appears to be the result of the distribution of wetland swales on the landscape, its dispersal ability, and water table fluctuations in response to climatic conditions. Excessive drought or flooding can cause local population decline and extinction (Sieg and King 1995). Available data (discussed later in this chapter) suggest that the presence of seeds at sites with adequate moisture allows the reestablishment of those local populations. In this manner, the metapopulation shifts in time and space in response to the dynamics of the water table. In addition, the orchid is threatened by habitat conversions such as plowing on nonpublic land and by ill-timed mowing, grazing, burning, water table manipulation, and other management activities on both public and private lands (U.S. Fish and Wildlife Service 1996). When it is possible to protect orchid habitat from these activities and uses, the problem of locating that protection over time remains. Capturing the population dynamics of a complex plant such as the orchid in an optimization model is challenging, however, so this is an exploratory model. We begin by summarizing the population dynamics we want to capture and then proceed to the model formulations.

The orchid is a perennial plant characterized by erratic above-ground growth and flowering. Periods of high numbers, usually linked with above-average precipitation, are followed by years when the orchids have seemingly disappeared (Bowles et al. 1992). The orchid has three distinct stages that we will model: seeds, protocorms, and above-ground plants. Little is known about seed dispersal in this species. Orchid seeds may be wind-dispersed (Bowles 1983), but their minute size, air-filled testa, water-repellent lipoid layer, and buoyancy make them well-equipped for water dispersal (Rasmussen 1995). In our model, we assume that the seeds disperse exclusively by water (flooding). Thus seed dispersal, carrying capacity, seed production, survivability, and other population dynamics of the orchid are all strongly influenced by yearly climate condition. We define discrete climate conditions and then optimize the placement of orchid habitat based on expected values across those climate conditions; some explanation is necessary before we proceed.

Information on the population dynamics of the orchid is limited, but perhaps the most telling observation occurred recently on the Sheyenne National Grassland. The orchid recovered dramatically beginning in 1992 (a wet year) after 5 relatively dry years in which above-ground orchid populations had dwindled to very low numbers (Sieg and King 1995). It is unlikely that this recovery can be attributed to plants returning from dormancy because recent demographic data collected on the Sheyenne National Grassland indicate that most above-ground plants live 3 years or less, and once they are absent, the odds of them remaining absent the next year are 80% or more (Sieg and King 1995). An alternate explanation for this recovery is the presence of a seed bank with at least some viable seeds persisting through the drought years. From this observation, we conclude that the orchid seeds (and thus the orchids themselves) do not seem to be subject to catastrophic loss from a single or even a string of dry years. The effects of seed production and dispersal last long enough (in terms of their viability) that we can expect the wet years to compensate for dry years. By this logic, we would be more concerned with maximizing the long-term mean population levels under protection and less concerned with yearly population fluctuations as long as we take the viability and dispersal of the seed bank into account.

We make the following assumptions about the orchid population dynamics with each year defined as the calendar year from January to December.

- Seeds that have accumulated are subject to transport, regardless of age, during wetter years when flooding occurs.
- Seeds are not produced until late in the year (September) and do not germinate until, at the earliest, the next spring.
- Germinated seeds spend a year underground as protocorms and emerge the next year (with a given success rate) as above-ground plants.

- A predictable portion of the above-ground plants flower each year, yielding a constant reproductive (seed production) rate per plant.
- Seeds remain viable at a decreasing rate (fixed percentage loss per year) over a number of years.
- During dry years, no seed transport occurs at the scale of our model.
- During wetter years, when flooding occurs, seed transportation takes place in the spring, before the production of that year's seeds. Germination takes place after transport, but again this applies only to seeds produced in previous years.
- Different areas have different orchid carrying capacities depending on yearly climate, determined mostly by topography.
- Seed production, seed viability, germination rates, protocorm emergence rates, and plant survival rates are functions of the climate condition in the year the rates are applied, not climate conditions in previous years. We recognize that this simplifying assumption may ignore the influence of climate condition in late summer on the formation of the perennating bud.

The landscape where the orchid appears typically is rolling grassland with swales and hummocks as the primary distinguishable topographic features. Orchid habitat in different climate conditions and seed dispersal through different flooding conditions are determined largely by this swale—hummock topography, so we characterize the spatial variables in our model by swale complexes. Each complex includes multiple contours: areas of different elevation that surround the deepest contour that is the lowland of the given swale.

Little is known about the actual seed dispersal process during the wetter years that produce flooding. In the case example, we investigate two sets of assumptions that bracket the possibilities that seem reasonable. It is assumed that no seeds leave or enter the given study area from the outside, creating a closed system for modeling purposes (but other assumptions could be accommodated).

The Model

We define the following indexes, variables, and parameters for use throughout the chapter:

Indexes

i and h index swale complexes,
j and k index contours within a swale complex,

t indexes time (in years),

w indexes climate conditions.

Variables

S_{ijt} = the expected above-ground plant population in swale i, contour j, and time t,

D_{ijt} = the expected number of viable seeds in swale i, contour j, and time period t,

B_{ijt} = the expected protocorm plant population in swale i, contour j, and time t,

Y_{ijt} = the area in swale i and contour j, protected for orchids in time period t.

Parameters

N_{ij} = the initial value for S_{ijt},

M_{ij} = the initial value for D_{ijt},

L_{ij} = the initial value for B_{ijt},

p_w = the probability of climate w occurring in any given year,

v_{ijw} = the proportion of seeds that do not germinate but remain viable for possible future germination in swale i, contour j, climate w in any given year,

g_{ijw} = the proportion of seeds that germinate in swale i, contour j, climate w in any given year,

r_{ijw} = the number of seeds produced per above-ground plant in swale i, contour j, climate w in any given year,

u_{ijhkw} = the proportion of seeds from swale h and contour k that, when connected to swale i and contour j by climate condition w, will settle in swale i and contour j,

q_{ijw} = the success rate of germinated seeds becoming protocorms in swale i, contour j, and climate w in any given year,

f_{ijw} = the rate at which protocorms survive and emerge as above-ground plants in swale i, contour j, and climate w from one year to the next,

z_{ijw} = the survival rate of above-ground plants in swale i, contour j, and climate w from one year to the next,

c_{ijw} = the above-ground plant carrying capacity in swale i, contour j, and climate w in any given year,

A_{ij} = the area in swale i, contour j,

$\bar{Y}$ = a policy-driven limit on the total amount of orchid habitat that can be protected for orchids,

a_{ijw} = the protocorm carrying capacity in swale i and contour j and climate w in any given year.

General Formulation

To capture the orchid population dynamics, we begin by setting initial conditions for the above-ground population, viable seed numbers, and protocorm population:

$$S_{ij0} = N_{ij} \qquad \forall i, \forall j, \tag{9.1}$$

$$D_{ij0} = M_{ij} \qquad \forall i, \forall j, \tag{9.2}$$

$$B_{ij0} = L_{ij} \qquad \forall i, \forall j. \tag{9.3}$$

The viable seed numbers in subsequent years are determined by

$$D_{ijt} = \sum_w p_w \left\{ \sum_h \sum_k e_{ijhkw} \left[v_{hkw} \times D_{hk(t-1)} \right. \right. \tag{9.4}$$

$$\left. \left. + r_{hkw} \times S_{hk(t-1)} \right] \right\} \quad \forall i, \forall j, t = 1, \dots, T.$$

Note:

e_{ijhkw} = 1 if swale $h = i$, contour $k = j$, and contour j is not flooded by climate w,

r = u_{ijhkw} if climate w connects swale i and contour j with swale h and contour k,

r = 0 otherwise.

Equation (9.4) determines the viable seed numbers in all subsequent years, based on the previous year's population (and the seed production from it) and the surviving viability of previously produced seeds. The different climate conditions and dispersal occurring in them are accounted for probabilistically in equation (9.4). The e_{ijhkw} coefficients are defined to correctly account for all seeds in any given time period.

The protocorm populations in subsequent years are determined by

$$B_{ijt} \leq \sum_w p_w \left[(q_{ijw} g_{ijw}) \times D_{ij(t-1)} \right] \quad \forall i, \forall j, t = 1, \dots, T, \tag{9.5}$$

$$B_{ijt} \leq \sum_w p_w (a_{ijw} A_{ij}) \quad \forall i, \forall j, t = 1, \dots, T. \tag{9.6}$$

The protocorm population in each area, in each time period after the initial one, is determined by whichever of (9.5) or (9.6) is binding (is the limiting factor). Inequality (9.5) limits the protocorm population by the germination and success rate of seeds becoming protocorms. Inequality (9.6) limits the protocorm population by the space limits (carrying capacity) of the habitat area.

The above-ground populations in subsequent years are determined by

$$S_{ijt} \le \sum_w p_w \left[f_{ijw} \times B_{ij(t-1)} + z_{ijw} S_{ij(t-1)} \right] \quad \forall i, \forall j, t = 1, \dots, T, \quad (9.7)$$

$$S_{ijt} \le \sum_w p_w \left(c_{ijw} Y_{ijt} \right) \quad \forall i, \forall j, t = 1, \dots, T. \tag{9.8}$$

The above-ground population is determined by whichever of (9.7) or (9.8) is limiting. Inequality (9.7) limits the above-ground population by the survival rate of existing plants and the emergence rate of the protocorms becoming above-ground plants. Inequality (9.8) limits the above-ground population by the carrying capacity of the habitat area.

In addition, we include

$$Y_{ijt} \le A_{ij} \quad \forall i, \forall j, t = 1, \dots, T, \tag{9.9}$$

$$\sum_i \sum_j Y_{ijt} \le \overline{Y} \quad t = 1, \dots, T. \tag{9.10}$$

Inequality (9.9) limits the areas allocated to orchid habitat to the areas available. Inequality (9.10) limits the total amount of orchid habitat allowed, a policy constraint.

Specific Formulation

In the model we built to demonstrate these formulations, we had to make some specific assumptions. First, we assumed that the v, r, g, q, f, a, z, and c parameters vary only by the contour and climate conditions but not individual swale complexes, thus allowing us to drop the i subscript from those parameters.

Second, we assumed that we only have three discrete climate conditions: a "dry" condition (indexed 1) that represents drought or near-drought conditions, a "mid" condition (indexed 2) that allows some water-based dispersal of seeds, and a "wet" condition (indexed 3) that disperses seeds with flooding out to the limits of the orchid habitat. We also assume three contours for each swale complex: a "deep" contour (indexed 1) that delineates the orchid habitat during the dry climate condition, a "medium" contour (indexed 2) that delineates the extent of seed dispersal in the mid climate condition, and a "high" contour (indexed 3) that delineates the extent of the area that is potential orchid habitat under any climate condition. During the dry climate condition, again, only the deep areas have any appreciable orchid habitat. During the mid climate condition, the deep and medium contours are good orchid habitat, but diminished habitat carrying capacity occurs in the high contour. During the wet climate condition, the carrying capacity in the deep contour is diminished by deep flooding, but the medium and high contours are good orchid habitat.

Third, we assumed (initially) that seed dispersal over whatever area is flooded by a given climate condition is random, mixing all seeds and placing them uniformly throughout the affected areas when the waters recede and the seeds settle out on the ground. This preserves the correct number of seeds and allows us to calculate expected seed dispersal coefficients based on area alone. A formulation for these specific assumptions is as follows:

Maximize

$$\sum_i \sum_{j=1}^{3} \sum_t S_{ijt}, \tag{9.11}$$

subject to

$$S_{ij0} = N_{ij} \quad \forall i, \forall j, \tag{9.12}$$

$$D_{ij0} = M_{ij} \quad \forall i, \forall j, \tag{9.13}$$

$$B_{ij0} = L_{ij} \quad \forall i, \forall j, \tag{9.14}$$

$$D_{ijt} = p_1(v_{j1} \times D_{ij(t-1)} + r_{j1} \times S_{ij(t-1)}) \tag{9.15}$$

$$+ p_2 \left[\sum_h \sum_k e_{ijhk2} \left(v_{k2} \times D_{hk(t-1)} + r_{k2} \times S_{hk(t-1)} \right) \right]$$

$$+ p_3 \left[\sum_h \sum_k \frac{A_{ij}}{\sum_{h'} \sum_{k'} A_{h'k'}} \left(v_{k3} \times D_{hk(t-1)} + r_{k3} \times S_{hk(t-1)} \right) \right]$$

$$\forall i, \forall j, t = 1, \ldots, T,$$

$$e_{ijhk2} = X_{ih} \frac{A_{ij}}{\sum_{h'} X_{ih'} \sum_{k'=1}^{2} A_{h',K'}} \quad \text{for } j = 1, 2 \text{ and } k = 1, 2$$

$$= 1 \qquad\qquad \text{for } j = 3, k = 3, i = h$$

$$= 0 \qquad\qquad \text{otherwise,}$$

$$X_{ih} = 1 \text{ if the \#2 contours of swale } i \text{ and } h \text{ are connected}$$
$$\text{by condition 2, including when } i = h, 0 \text{ otherwise,}$$

$$B_{ijt} \le \sum_w p_w \left[(q_{jw} \times g_{jw}) \times D_{ij(t-1)} \right] \quad \forall i, \forall j, t = 1, \ldots, T, \quad (9.16)$$

$$B_{ijt} \le \sum_w p_w (a_{jw} A_{ij}) \quad \forall i, \forall j, t = 1, \ldots, T, \quad (9.17)$$

$$S_{ijt} \le \sum_w p_w \left[f_{jw} \times B_{ij,(t-1)} + z_{jw} S_{ij(t-1)} \right] \quad \forall i, \forall j, t = 1, \ldots, T, \quad (9.18)$$

$$S_{ijt} \le \sum_w p_w (C_{jw} Y_{ijt}) \quad \forall i, \forall j, t = 1, \ldots, T, \quad (9.19)$$

$$Y_{ijt} \le A_{ij} \quad \forall i, \forall j, t = 1, \ldots, T, \quad (9.20)$$

$$\sum_i \sum_j Y_{ijt} \le \overline{Y} \quad t = 1, \ldots, T. \quad (9.21)$$

All variables and parameters (other than X_{ih}) are defined as before, except for the omission of the i subscript from the parameters previously indicated. The objective function (9.11) maximizes the expected above-ground orchid population totaled across all habitat areas and time periods. All time periods are included to reflect a concern for all years in the planning time horizon (including a recovery process), but other objective functions could also be used.

To bracket the reasonable possibilities for seed dispersal under flooding conditions, we also examine a more conservative set of dispersal assumptions in equation set (9.15). This alternative is formulated by setting all

$$X_{ih} = 1 \text{ if } i = h \quad (9.22)$$
$$= 0 \text{ otherwise}$$

and by replacing the third term (for climate condition 3) in (9.15) with

$$p_3 \left[\sum_h \sum_k \frac{A_{ij}}{\sum_{j'} A_{ij'}} \lambda_{ih} \left(V_{k3} \times D_{hk(t-1)} + r_{k3} \times S_{hk(t-1)} \right) \right], \quad (9.23)$$

where

$$\lambda = \frac{\gamma_{ih}}{\phi_h} \text{ if } i \ne h \quad (9.24)$$

$$= 1 - \left(\sum_{h' \ne i} \alpha \frac{\gamma_{ih'}}{\phi_i} \right) \text{ if } i = h$$

γ_{ih}, $\gamma_{ih'}$ = the length of the common border between swale complex i
  and swale complex h (or h'),

ϕ_i, ϕ_h = the perimeter of the (source) swale complex i (or h),

α = the proportion of seeds that would exit swale complex h in climate condition 3 if there were no impediments anywhere around
  its perimeter.

Equation set (9.22) assumes that under climate condition 2, no interswale complex seed dispersal takes place; only seeds within each swale complex are mixed across the contours contained in that swale complex. In (9.23) and (9.24), under climate condition 3, seeds are exchanged only between adjacent swale complexes. These exchanges take place as a linear function of the ratio of common perimeter to total perimeter (of the source cell). They are also scaled by the α parameter, which indicates the proportion of seeds that would disperse if the entire perimeter of the source swale complex were common with some other swale complex (as opposed to nonhabitat areas including areas outside the study area boundary). This formulation assumes a tendency of the seeds to disperse in a 360% pattern, with seeds staying in the source swale complex if hindered by the topography of the nonhabitat areas (hum-mocks). All seeds within each swale complex, including those that enter from an adjacent swale complex, are again randomly mixed across the relevant con-tours under climate condition 3. If it was desired, the α parameter could vary according to the source swale complex to account for source swale size or other special properties, but we assume it to be a constant.

Landscape

Plate 13a shows an enhanced graphic of the selected study area, a 300-ha square located in the Sheyenne National Grassland in southeastern North Dakota. The topographic data were obtained from the U.S. Geological Sur-vey Earth Resources Observation Systems Data Center in Sioux Falls, South Dakota (U.S. GeoData Digital Elevation data). These data classify 7.5-minute land units into discrete elevation classes defined in feet, so we retain those units. On the basis of this landscape and observed flood conditions, the three contours were identified by the color codes indicated. The deepest areas (con-tour 1) are coded blue, the middle ground (contour 2) are coded light green, and the highest habitat areas (contour 3) are coded dark green. The yellow and orange areas are hummocks and are designated as nonhabitat. This graphic was also used to delineate 22 swale complexes, as shown in plate 13b. As is evident, several of the swale complexes do not have the low contour, but all have the other two. With the initial seed dispersal assumptions, the third (wettest) climate condition floods and connects all the area contained by the dark green contour; also, the second climate condition connects the middle (light green) contour areas for swales 1–3, swales 13–16, and swales 20 and 21. With the conservative seed dispersal assumptions, the wettest climate condition connects only the adjacent swale complexes, and the mid climate condition mixes only the seeds within each swale complex.

Parameters

For demonstration purposes, we wanted to model a situation in which past management practices have created a loss in orchid populations, and the modeling effort is used as an aid in developing a strategy to identify critical habitat for protection. Thus we assumed initial presence of above-ground plants and protocorms to be limited to swale complex 2. We assumed the initial density of above-ground plants in this swale complex to be one-half the average carrying capacity (across climate conditions). We assumed the initial density of protocorms to be one-fourth the average protocorm carrying capacity, and we assumed that initial seed density was 10,000 seeds/ha in swale complex 2 and 100 seeds/ha for all other habitat areas.

The climate condition probabilities were difficult to estimate because the relationship between climate indicators such as precipitation and on-the-ground conditions is lagged and indirect. We were able to obtain unpublished ground water (well) data from the U.S. Geological Survey for two wells in the area for the years 1976–1997. These ground water data are more directly related to orchid habitat conditions than are precipitation records, but 21 years obviously is inadequate to provide a statistically valid estimate of the desired probabilities. However, these data do provide a rough estimate for use in our demonstrative model. The data from one of the wells indicated 8 years (1976, 1979, 1984, 1986, 1993–1996) that we could associate with wet years in terms of orchid habitat, where the ground water peaked within 2½ feet of the surface. Treating these as the wet years in the historical data suggests a probability of 8/21 (.381) for condition 3. From plate 13a, the middle contour is about 3 feet deep, so we counted the years with peak ground water depths between 2½ and 5½ feet as years with climate condition 2; there were seven of them, implying a probability of 7/21 (.333). The remaining years, where the well's ground water depth did not get within 5½ feet of the surface, were treated as climate condition 1 years, with a probability of 6/21 (.286). The data from the second well were similar, except that they did not indicate 1984 as a wet year. To demonstrate the sensitivity of the model to these probabilities, we tried this adjustment, discussed further in the Results section.

The other parameter values are given in table 9.1. These parameters, under

TABLE 9.1

Parameter Values for Demonstration Model

	Contour								
	Deep			Moderate			Shallow		
	Wet	Mid	Dry	Wet	Mid	Dry	Wet	Mid	Dry
v_{jw}	.5	.6	.7	.5	.6	.7	.5	.6	.7
g_{jw}	.01	.11	.05	.11	.07	.01	.11	.01	0
r_{jw}	0	5902	4864	9923	2529	0	6615	0	0
q_{jw}	0	.13	.06	.13	.05	0	.01	0	0
f_{jw}	0	.09	.08	.09	.07	.01	.09	0	0
z_{jw}	0	.5	.25	.23	.78	0	.23	.09	0
c_{jw}	0	5.46	.2	5.46	.2	0	1.0	.54	0
a_{jw}	0	109	4	109	4	2	20	11	0

the various scenarios, are based on published and unpublished data collected on the Sheyenne National Grassland beginning in 1987 (see Sieg and King 1995; Hof et al. 1999). The dimensions of the case example create a linear program with approximately 8000 constraints and 8200 variables.

Results

Initial Seed Dispersal Assumptions

Protecting habitat for the orchid is likely to be difficult, with limited resources available. Therefore the typical problem situation is to optimally locate a limited amount of habitat over time. Plate 14 presents the management strategy suggested by the model with the initial seed dispersal assumptions and 50 ha of habitat allowed to be protected (see also table 9.2, column 2). This solution and the ones that follow strongly suggest that much higher priority should be given to the two lower contours for habitat protection. It is largely within those two contours that the spatial and temporal allocation takes place. The management strategy indicated by the model in this solution has three distinct phases. Initially, habitat is protected around swale complex 2 to assist the fledgling population in establishing itself. Habitat protection gradually increases elsewhere until, by year 16 (depicted in plate 14a), a nearly even level

TABLE 9.2

Expected Values of Orchid Populations from Model Solutions
with the Initial Seed Dispersal Assumptions

Year	Base Solution	Even Allocation	Even Allocation for Contours 1 and 2
1	2.37	2.37	2.37
2	2.02	1.06	1.85
3	6.79	3.65	5.38
4	9.04	6.75	8.31
5	8.08	6.52	7.59
6	10.75	7.34	9.37
7	15.10	10.64	13.39
8	17.56	13.06	15.88
9	21.30	15.34	18.93
10	27.74	19.72	24.50
11	34.53	24.36	30.67
12	42.40	28.26	37.51
13	53.43	31.39	47.00
14	67.11	32.03	58.94
15	83.42	32.36	73.08
16	101.30	32.69	87.94
17	104.60	32.94	99.40
18	105.91	33.10	103.18
19	106.30	33.21	103.27
20	106.30	33.29	103.27
21	106.30	33.34	103.27
22	106.30	33.37	103.27
23	106.30	33.39	103.27
24	106.30	33.41	103.27
25	106.30	33.42	103.27
Total	1457.55	567.01	1368.18

of habitat protection is placed in all contour 1 and 2 areas (except around swale complex 2, which is still fully protected). Then the solution begins concentrating on the contour 2 areas until, by year 20 (depicted in plate 14b), the solution reaches its maximum population, almost completely concentrating on the contour 2 areas. We also mapped the orchid population response to this habitat protection strategy, and, not surprisingly, the population closely follows the pattern of protected habitat. The first phase of this strategy clearly depends on the initial conditions. The second phase appears to include the lowest contour with the middle contour to build a seed bank (it is only in the lowest contour that seeds are produced in the dry climate condition; see table 9.1). Once a seed bank is established, the solution focuses on the middle contour areas, implicitly relying on the seed bank during dry years. The sensitivity of this strategy to the climate probabilities used and the importance of applying such a strategy in an adaptive management context are discussed later in this chapter.

The spatial and temporal layout of protected habitat, especially with regard to the two lower contours, is indicated to be important. Table 9.2 also presents solution population results with two limits on spatial and temporal flexibility. The third column in table 9.2 shows a solution in which the proportion of each swale contour area protected as habitat is constrained to be the same. This solution represents a simple-minded spatial allocation that simply treats all areas the same (proportionately). This solution indicates a very large loss in orchid population relative to the one in the second column of table 9.2: from 1458 to 567 plants over 25 years. However, most of this loss can be attributed to the allocation to the highest contour at the expense of the lower two. If it was known that the lower two contours should be given priority, the solution in the fourth column of table 9.2 might be a spatially neutral approach. Here all 50 ha of habitat are allocated to the lower two contours, but each swale contour area within those two contours is allocated an equal proportion of habitat. With this approach, the loss in orchid population is much smaller: about 89 plants over 25 years. Most of this loss is observed in time periods 11–16, with a 12.5% reduction. The maximum population level in the fourth column in table 9.2 is only about 3% lower than the maximum reached in the second column. Thus the spatial placement of habitat in the lower two contours is indicated to be more important for the middle years of a recovery period than for the maximum that is eventually reached. We investigate the robustness of this conclusion further.

Conservative Seed Dispersal Assumptions

Plate 15 and column 2 of table 9.3 present the optimized solution for the model with the more conservative seed dispersal assumptions and, again, 50 ha of protected habitat allowed. Priority is still indicated to be much higher for the lower two contour areas than for the highest contour area. Also note that the approach to the maximum population level is much slower (as expected) with these conservative seed dispersal assumptions. However, the basic strategy suggested by this solution is quite similar to the one in plate 14. The initial phase again protects orchid habitat around the initial population. By year 20

TABLE 9.3

Expected Values of Orchid Populations from Model Solutions
with the Conservative Seed Dispersal Assumptions

Year	Base Solution	Even Allocation	Even Allocations for Contours 1 and 2	Altered Climate Probabilities
1	2.37	2.37	2.37	2.19
2	2.02	1.06	1.85	1.88
3	6.79	3.65	5.38	6.30
4	7.71	4.62	6.40	7.08
5	7.83	4.41	6.55	7.10
6	10.48	6.05	8.75	9.59
7	13.70	8.33	11.52	12.56
8	16.23	9.84	13.69	14.84
9	20.35	12.26	17.13	18.86
10	26.40	14.92	22.21	24.92
11	33.22	16.45	27.95	31.75
12	41.81	18.32	35.14	40.81
13	50.14	21.07	40.03	48.08
14	56.78	24.00	44.65	54.35
15	61.47	27.43	48.51	62.18
16	67.61	29.86	53.65	69.92
17	75.04	30.90	59.56	79.53
18	83.16	31.43	66.51	91.07
19	92.99	31.83	75.31	99.15
20	101.54	32.17	85.90	99.31
21	104.11	32.41	93.63	99.46
22	104.98	32.60	96.48	99.50
23	105.34	32.72	98.91	99.51
24	105.66	32.79	100.06	99.52
25	105.81	32.84	100.62	99.53
Total	1303.54	494.33	1122.76	1278.99

(depicted in plate 15a), a substantial amount of habitat in all contour 1 and 2 areas is protected, but the spatial effect of the initial conditions obviously remains much longer than in plate 14a because of the slower and more difficult expansion of the orchid implied by the conservative seed dispersal assumptions. By year 25, the solution has still not reached a maximum population level (column 2, table 9.3), and further expanding the time horizon created a model that we were not able to solve. However, the allocation in plate

15b is quite similar to plate 14b. As before, the orchid population associated with this solution closely follows the pattern of habitat protection.

In column 3 of table 9.3, we present a solution similar to column 3 of table 9.2, where the proportion of each swale-contour area protected as habitat is constrained to be the same. Results are similar to those in table 9.2, with a very large loss in orchid population. Again, this is largely attributable to the allocation of habitat to the contour 3 areas at the expense of the contour 1 and 2 areas.

In column 4 of table 9.3, a solution similar to column 4 in table 9.2 is presented, where all 50 ha are allocated to the lower two contours but with an equal proportion protected in each swale contour area within those two contours. The loss in orchid population is about 19.5% for years 13–20. The basic pattern is quite similar to that in table 9.2, but the more conservative dispersal assumptions appear to make the spatial allocation in the middle years of recovery a bit more important than before. The robustness of this basic solution strategy to significantly different seed dispersal assumptions is noteworthy, but we found that the third phase of this strategy was highly sensitive to the climate condition probabilities.

Recall that in the ground water height data from the second well that we were able to obtain, the year 1984 was not indicated to be a wet year, as it was in the first well's data. Making this single adjustment to the estimates of climate condition reverses the probabilities of the middle and wet climate conditions (now .381 for climate condition 2 and .333 for climate condition 3). Plate 16 and column 5 of table 9.3 depict the solution of the model that results from making this adjustment with the conservative seed dispersal assumptions (the initial model responded similarly to this change in climate condition probabilities). The first two phases of the solution strategy with these altered probabilities are similar to those just discussed, as evidenced by the similarity in plates 15a and 16a (year 20 in both cases). By year 25, however, the third phase strategy in plate 16b (see also column 5, table 9.3) is quite different. Now, the lowest contour areas are completely protected, with ancillary middle contour areas included as possible given the 50-ha total limit. The altered probabilities are a bit "drier," based on one less historical year being considered in condition 3 (and thus one more in condition 2). With these slightly drier probabilities, it is now preferable to focus on the lowest contours that provide habitat in the driest years, even after a large seed bank is established.

The sensitivity to the climate probabilities points out the particular importance of using a model such this one (where risk is handled with fixed probabilities and optimization is based on expected values) in an adaptive management process (Walters 1986) as discussed in chapter 1. Clearly, the tenability of such a model erodes rapidly with time as variances around expected values increase. These weaknesses would be ameliorated if the status of the system being modeled was monitored regularly and the model was rerun with new initial conditions as they became known. In such a process, if circumstances never allowed the orchid to recover very far (e.g., because of a string of dry years), the latter two phases of the management strategies described earlier might never be implemented. If events proceed close to the expected value predictions, then the calculated strategy might remain tenable much longer. Conversely, if the third phase was reached but became untenable because of a random or catastrophic event, the adaptive process would reset the management strategy to an initial phase based on the new conditions. No clearcut strategy to managing risk is ever available, but the analysis in this chapter suggests that arbitrary timing and placement of orchid habitat protection will not be as effective as a more calculated approach. In chapter 10, we return to animal populations to consider the effects of a species needing resources from two or more habitat types.

10

HABITAT EDGE EFFECTS

For decades after Aldo Leopold's publication of *Game Management* (1933), wildlife habitat management efforts tended to emphasize creating induced edge (Thomas et al. 1979) and coverts (where three or more habitat types come together) because of the positive effects observed for many wildlife species (Hunter 1990). Although Leopold (1933) recognized that not all species respond positively to habitat edges, the emphasis on creating edge persisted at least into the 1970s (Giles 1978). By the 1970s and 1980s, however, negative edge effects on habitat interior species were becoming recognized (e.g., Soulé 1986) and were of increasing concern in forest management (e.g., Harris 1984). Although a number of geometric methods for optimizing edge configurations in forest tracts have been investigated (e.g., Conlin and Giles 1973; Harris and McElveen 1981), these "mechanical" approaches rely on arbitrary thresholds of differences between adjacent areas to define edge. Representing the variety of ecological processes underlying different population responses to habitat with such methods is difficult at best.

The purpose of this chapter is to explore an ecological process-based modeling approach to optimizing edge and interior habitat configurations resulting from forest treatments. Hunter (1990) provides a useful synopsis of edge habitats and wildlife population response classes from which we begin. His Group A wildlife species are associated with edges because they need resources found separately in two distinct habitat types. An example would be

This chapter was adapted from M. Bevers and J. Hof, Spatially optimizing wildlife edge effects in forest management linear and mixed-integer programs, *Forest Science* 45, no. 2 (1999): 249–258, with permission from the publisher, The Society of American Foresters.

a species whose nesting or denning requirements are found in dense vegetation but whose foraging occurs in sparsely vegetated openings or in an altogether different vegetation type. Where two habitat types adjoin, an ecotone generally occurs. Whereas some habitat edges are abrupt, with very small ecotones, other transitions are more gradual, forming large ecotones. Hunter's group B wildlife species need an ecotone habitat and tend to be absent from abrupt edges. Because their populations usually do not extend into either habitat interior, they can be viewed as a single-habitat specialist whose habitat is the ecotone. His Group C wildlife species are associated with one or the other habitat type, extending into ecotone transition zones. Additionally, other wildlife might be interior species for which the ecotone is inhospitable or habitat generalists that can meet all their needs within either habitat type, including the ecotone. Extending these classes to species with three or more habitat requirements would be straightforward.

We extend the previous chapters in this section, which focused on species in Group C with a single habitat requirement, to Group A, for which edge is important because of the species' requirements for two or more resources (ecotone considerations are briefly discussed later in this chapter). As Orians and Wittenberger (1991:S32) put it, "Good foraging habitats may provide poor escape cover and low-quality nest sites, and vice versa. Some of these problems can be reduced if the organism has good mobility, but there is always a cost entailed by exploiting resources at a distance from home base. . . . The location of the resource with the most restricted distribution exerts the most powerful constraint on where individuals settle" (see also Orians and Pearson 1979; Schoener 1979).

Using the nest site and forage requirement example, Group A species might be limited by the amount of and the distance between forage from one habitat type and nest sites in another. In addition to being limited by resource availability, populations are also limited by the ability of their organisms to successfully reproduce and disperse to suitable sites. Reaction–diffusion theory (Skellam 1951; Levin 1974; Okubo 1980; Holmes et al. 1994) and the discrete reaction–diffusion model (Levin 1974; Allen 1987) again provide the framework for modeling populations limited by reproduction and dispersal processes in this chapter.

Formulation

To capture edge effects on populations, we formulate a linear programming model based on reproduction and dispersal, nest site capacity, forage production, and effective forage consumption imposed as limiting factors. The amount of nest site capacity and forage production result from resource management activities within each land unit. As in previous chapters, these management activities are modeled with decision variables that represent different vegetation treatment schedules applied over an entire planning horizon to all or a portion of each land unit, as selected by the linear program. Optimization is used to simultaneously assign site-specific forage production to nest locations for consumption while recognizing the inefficiencies of foraging far from nest sites. In this chapter, we also incorporate a MAXMIN operator from Bevers et al. (1995) that iteratively emphasizes the time periods with the least population within the planning time horizon. The linear programming model is as follows:

Iteratively maximize

$$\lambda \qquad (10.1)$$

subject to

$$\sum_i S_{it} \geq \lambda \quad t = 1, \ldots, T \qquad (10.2)$$

and

$$S_{it} \leq \sum_j (1 + r_j) g_{ji} S_{j(t-1)} \quad \forall i \quad t = 1, \ldots, T, \qquad (10.3)$$

$$S_{i0} = N_i \quad \forall i, \qquad (10.4)$$

$$S_{it} \leq \sum_k c_{ikt} X_{ik} \quad \forall i \quad t = 1, \ldots, T, \qquad (10.5)$$

$$S_{it} \leq \sum_j d_{ji} Z_{jit} \quad \forall i \quad t = 1, \ldots, T, \qquad (10.6)$$

$$\sum_i Z_{jit} \leq \sum_k f_{jkt} X_{jk} \quad \forall j \quad t = 1, \ldots, T, \qquad (10.7)$$

$$\sum_k X_{ik} = A_i \quad \forall i, \qquad (10.8)$$

$$\sum_i \sum_k m_{ikpt} X_{ik} \geq M_{pt} \quad \forall p \quad t = 1, \ldots, T, \qquad (10.9)$$

where

i and j both index all of the land units (e.g., forest stands) in the model,

t indexes discrete time periods from initial conditions at $t = 0$ to the end of the planning horizon at $t = T$,

k indexes the set of forest treatment schedules under consideration for managing the forest cover in each land unit throughout the planning horizon,

p indexes the set of applicable policy constraints (if any),

S_{it} is a state variable representing the wildlife species population (nesting adult pairs in our examples) in land unit i at time t (S_{it} and S_{jt} denote the same variable when $i = j$),

λ is a variable representing the smallest periodic total population from among the time periods whose populations are being optimized at any given objective function iteration (as discussed later in this chapter),

X_{ik} is a management decision variable representing the area in land unit i to be managed under treatment schedule k throughout the planning horizon (X_{ik} and X_{jk} denote the same variable when $i = j$),

Z_{jit} is a forage allocation variable describing the amount of forage produced in land unit j that is made available for consumption by organisms nesting in land unit i during time period t,

r_j is a parameter representing a net per capita periodic reproduction rate (not accounting for dispersal mortality, as discussed later in this chapter),

g_{ji} is a parameter representing the proportion of organisms (both adults and fledgling offspring) in land unit j that are expected to disperse to land unit i each time period,

N_i is a parameter describing the initial population in land unit i at the start of the planning horizon,

c_{ikt} is a parameter describing the nest site capacity produced in time period t per unit of area in land unit i that is managed according to treatment schedule k,

d_{ji} is a parameter describing the utilization efficiency (declining with distance) of forage produced in land unit j and consumed by organisms nesting in land unit i,

f_{jkt} is a parameter describing the amount of forage produced in period t per unit of area in land unit j that is managed according to treatment schedule k,

A_i is a parameter defining the amount of area in land unit i,

M_{pt} is a parameter describing the least (or most, when equation 10.9 is used as a less-than-or-equal-to constraint) total amount of a given

resource (e.g., timber volume, net revenue, forage) to be produced
from the planning area in time period t under policy constraint p,
m_{ikpt} is a parameter describing the amount of resource production (see
M_{pt}) under policy constraint p expected in time period t per unit of
area in land unit i that is managed under treatment schedule k.

With this formulation, any objective function can be used that tends to
maximize population so that at least one ecological factor (represented by
equations 10.3, 10.5, and 10.6) limits the population (S_{it}) in each land unit and
time period. Equations (10.1) and (10.2) are used here to iteratively maximize
each periodic population total, beginning with the lowest and continuing to
the next lowest and so forth throughout the planning horizon. After each iteration, the population totals for the periods constraining λ are used in subsequent iterations as right-hand side values in equation (10.2), replacing λ for
those periods. This procedure is risk averse in that it maximizes the population during the time periods when it is the lowest and thus the most at risk
(again, see Bevers et al. 1995).

As discussed in the part II introduction and in previous chapters, the equations in (10.3) combined with initial conditions in equation (10.4) form a coupled map lattice (Kaneko 1993) based on reaction–diffusion constraints defining reproduction and dispersal rates for a population distributed over multiple
land units. The nesting adult population in each land unit for each time period
is limited by that unit's population in the preceding time period expanded by
the net reproduction rate r value (r), less the proportion that leave the home
unit each time period ($1 - g_{ii}$), plus the number of immigrants from other land
units. Generally, the sum of the dispersal proportions from a given land unit
is less than 1, reflecting dispersal mortality. We do not include any immigration from outside the planning area in equation (10.3), and organisms dispersing beyond the planning area are treated as lost from the population,
although other assumptions could also be modeled.

Equation (10.5) further limits the nesting adult population in each land unit
to the amount of nest site capacity produced in each time period by the combination of forest treatment schedules selected. This constraint is similar to those
used in chapters 7–9. In this chapter, we introduce positive edge effects by
including forage requirements, as modeled with equations (10.6) and (10.7).

Equation (10.6) restricts the population nesting in each land unit to the
amount of forage available in each time period from that unit and other nearby
land units. Many species forage or hunt well beyond the nest or den sites but
at some energetic cost. The farther from the nest site that organisms have to
go for food, the more time and energy they consume and the greater the risk
they incur, eventually reaching a distance beyond which the nest cannot be
maintained. Our forage utilization efficiency parameters (d_{ji}) decline expo-

nentially with distance between the forage source (j) and the nest site (i). This represents the effects of higher energetic demand by requiring a larger forage allocation from distant sources per nest site supported. At the point where d_{ji} is reduced to zero for a given forage source and nest site, no amount of forage is useful. Some species may forage only in the area immediately surrounding the nest site, with nesting and foraging activities remaining within a single land unit. In such cases, the population will be locally limited by whichever resource is most scarce, and a single carrying capacity constraint set similar to equation (10.5) is enough.

The forage produced in each time period by the combination of forest treatment schedules selected for a given land unit is partitioned among nest sites in that and nearby land units by equation (10.7). Because effective forage utilization declines with distance in equation (10.6), nearby forage allocations are favored, and forage allocations beyond the maximum foraging distance are prevented.

Equation (10.8) requires that all land in each land unit be allocated to one or more of the forest treatment schedules under consideration. Equation (10.9) might represent any number of policy constraints that could be imposed over any portion of the planning area. In total, then, the model is structured so that the forest management variables (X_{ik}) are selected to optimize the location of forage and nest site habitats over time to produce the best possible population response given limited resources and a simplified set of species reproduction, dispersal, and foraging traits.

Case Example

Because model results are easiest to see and interpret with high-contrast edges (defined by the degree of difference between the two habitats; see Hunter 1990), we use simple static models and a single type-conversion management option for our initial experiments. In the static models, equations (10.1) and (10.2) are replaced with an objective function that just maximizes total population:

Maximize

$$\sum_i S_i. \tag{10.10}$$

Likewise, the t subscript for time periods is dropped from equations (10.3), (10.5)–(7), and (10.9), and equation (10.4) is eliminated.

To test for edge effects, we constructed a series of models for a forest comprising a block of 81 initially mature stands represented by a 9×9 grid of square cells completely surrounded by nonforest. We use square stands here for convenience. Actual vegetation polygons could also be used. In our models, the area parameters (A_i) in equation (10.8) are set to one for each stand. Mature forest stands provide consistent nesting habitat ($c_{ikt} = 1$ population unit) with forage production (f_{jkt}) that varies for each experiment. The surrounding nonforest area supports no nest sites ($c_{ikt} = 0$) but provides a consistent forage supply ($f_{jkt} = 1$). For most of our models in this chapter, we assume a wildlife species for which a forest stand represents a large area, so that one population unit implies 100 breeding pairs. Other scale relationships are examined later in this chapter. We also assume a wildlife species that can forage up to two cells away from the nest site cell. Consequently, the model includes two concentric layers of nonforest cells surrounding the forest, forming a 13×13 grid overall. As with forest stands, different polygons could also be used to represent nonforest. Forage utilization efficiency parameters (d_{ji}) are set to decline exponentially over a total distance of two cells from any nest site cell, as shown in figure 10.1. Harvests can be used to change forage production within the forest to varying degrees in each experiment.

We begin with a static model of a species for which adequate forage occurs in the same habitat type as the nest sites so that nest sites are the limiting habitat factor in any given land unit. In subsequent static models, we reduce the amount of forage available from forest (nesting) habitat and examine the resulting effects on population distribution. For each of these models, forest management decision variables X_{i1} represent the proportion of each stand converted to nonforest through permanent removals. Forest management decision variables X_{i2} represent the proportion of each stand retained as forest to provide nest sites. No management options are considered for the nonforest land units, and only X_{i1} variables (fixed to 1.0 by equation 10.8) are included for those 88 cells in all of our models.

Forage production in unharvested forest stands (f_{j2}) is set to 1.5 in this first model. As in the surrounding nonforest, forage production in harvested stands (f_{j1}) is set to 1. Note that because utilization efficiency within the nest site cell itself (d_{ii}) is 0.80 (see figure 10.1), the forage in an unharvested forest stand can support 1.2 population units (0.80 times 1.5 population units of available forage), more than can nest there. All net reproduction rate r values (r_j in equation 10.3) are set to 0.75, and all dispersal proportions (g_{ji}) are set to represent an exponential decline in dispersal probability with distance from the natal cell (as in Fahrig 1992), with a mean distance of 3.0 land units (see figure 10.2).

Results

Figure 10.3 shows the nesting population distribution indicated by this model in the absence of any policy constraints (note that only the 9×9 grid of for-

.01	.04	.06	.04	.01
.04	.11	.18	.11	.04
.06	.18	.80	.18	.06
.04	.11	.18	.11	.04
.01	.04	.06	.04	.01

FIGURE 10.1

Forage utilization efficiency parameter (d_{ji}) values defined over a maximum foraging distance of two cells, with the center cell in the figure designated as the nest site cell (subscript i) and each of the 25 cells as a potential forage source (subscript j).

est cells is shown). The obvious optimal solution is to retain all forest stands, resulting in a population of 77.37 units, or 7737 breeding pairs. All interior stands are occupied at full capacity (limited by equation 10.5), but nearly all stands around the edge of the forest are occupied at less than full capacity (limited by equation 10.3), with declining populations toward the corners. Because forage is plentiful at the nest sites, the surrounding nonforest has no positive influence on this population. In this model, the forest could even be removed from portions of the edge stands without reducing the wildlife population as long as enough forest is retained to support the nest sites indicated in figure 10.3 (although, in reality, this could disturb some species).

Similar models in which a single carrying capacity constraint set is limiting have been discussed in earlier chapters. We present this case as a reference point for the experiments that follow. Without any need to forage beyond the immediate vicinity of the nesting site, this species has a single habitat requirement most closely representing Hunter's Group C. As figure 10.3 shows, the population displays a negative edge effect such as might be expected for a habitat interior specialist.

Forage Constraint Effects

In our next model, we reduce the amount of forage available in the nesting habitat from 1.5 to .5 units per forest stand. Converting nesting habitat (with

.0014	.0020	.0027	.0033	.0035	.0033	.0027	.0020	.0014
.0020	.0031	.0045	.0059	.0066	.0059	.0045	.0031	.0020
.0027	.0045	.0074	.0115	.0139	.0115	.0074	.0045	.0027
.0033	.0059	.0115	.0244	.0401	.0244	.0115	.0059	.0033
.0035	.0066	.0139	.0401	.1704	.0401	.0139	.0066	.0035
.0033	.0059	.0115	.0244	.0401	.0244	.0115	.0059	.0033
.0027	.0045	.0074	.0115	.0139	.0115	.0074	.0045	.0027
.0020	.0031	.0045	.0059	.0066	.0059	.0045	.0031	.0020
.0014	.0020	.0027	.0033	.0035	.0033	.0027	.0020	.0014

FIGURE 10.2

Dispersal proportion parameter (g_{ji}) values defined with the center cell in the figure designated as the destination cell for the current breeding season (subscript i) and each of the 81 cells as a potential source cell from the preceding breeding season (subscript j).

some forage) to better foraging habitat by forest removal increases the available forage from .5 to 1 unit.

Figure 10.4a shows the effects of these changes on population distribution when no forest removals are allowed (as in figure 10.3), and figure 10.4b shows the population distribution under optimal management. Population totals for the two solutions are 34.75 and 47.24 units, respectively, suggesting that disregarding this species' foraging requirements by managing for it as an interior specialist could result in substantially reduced populations.

Figures 10.4a and 10.4b again display the smallest populations in the three cells forming each of the four corners of the forest (as in figure 10.3), limited by the reaction–diffusion constraints (equation 10.3). Unlike our previous model, however, both solutions also display larger populations in the remain-

.62	.81	.93	.99	1.0	.99	.93	.81	.62
.81	1.0	1.0	1.0	1.0	1.0	1.0	1.0	.81
.93	1.0	1.0	1.0	1.0	1.0	1.0	1.0	.93
.99	1.0	1.0	1.0	1.0	1.0	1.0	1.0	.99
1.0	1.0	1.0	1.0	1.0	1.0	1.0	1.0	1.0
.99	1.0	1.0	1.0	1.0	1.0	1.0	1.0	1.0
.93	1.0	1.0	1.0	1.0	1.0	1.0	1.0	.93
.81	1.0	1.0	1.0	1.0	1.0	1.0	1.0	.81
.62	.81	.93	.99	1.0	.99	.93	.81	.62

FIGURE 10.3

The optimal static equilibrium wildlife population distribution (in hundreds of breeding pairs) for a hypothetical species when forest nesting habitat also provides adequate forage within each stand.

ing cells of the outer two layers of forest stands, reflecting a pronounced positive edge effect. The interior of the forest no longer supports the largest cellular populations, with or without optimal management. This effect appears in response to the combination of limited forage within the forest and the abundance of forage outside the forest. The largest local populations occur one layer in from the edge near each of the corners, striking a balance between the negative effects of dispersal losses beyond the forest and the beneficial effects of proximity to forage in the forest corners and beyond. Although management that replaces surplus nest site capacity with better forage habitat within the forest does increase the overall population, the general pattern of edge response is little changed by these partial stand removals (compare figures 10.4a and 10.4b).

.28	.38	.43	.45	.46	.45	.43	.38	.28
.38	.50	.54	.46	.46	.46	.54	.50	.38
.43	.53	.40	.40	.40	.40	.40	.53	.43
.45	.46	.40	.40	.40	.40	.40	.46	.45
.46	.46	.40	.40	.40	.40	.40	.46	.46
.45	.46	.40	.40	.40	.40	.40	.46	.45
.43	.54	.40	.40	.40	.40	.40	.54	.43
.38	.50	.53	.46	.46	.46	.53	.50	.38
.28	.38	.43	.45	.46	.45	.43	.38	.28

FIGURE 10.4a

Static equilibrium wildlife population distributions (in hundreds of breeding pairs) for a hypothetical species when forest nesting habitat provides .5 population unit of forage within each stand given (a) no forest conversion and (b) optimal forest conversion to foraging habitat.

Next, we further reduce the amount of forage available in the nesting (forest) habitat from .5 unit per stand to zero. Forage is now available only in the area outside the forest (as before) or where forest is removed through management. Figure 10.5a shows the population distribution resulting from optimal management under this assumption. The population total is now further constrained by available forage to 36.74 units, but the optimal distribution pattern is little changed from our previous model (see figure 10.4). A positive edge effect still appears except in the corner area cells, with the interior of the forest slightly less populated than the edge. In this case, however, the interior forest population is maintained only through management, as figures 10.5b–10.5d show. To support the population in figure 10.5a, nearly 55% of the forest must be removed through widespread application of partial stand

.38	.51	.58	.61	.62	.61	.58	.51	.38
.51	.67	.67	.62	.61	.62	.67	.67	.51
.58	.67	.57	.57	.57	.57	.57	.67	.58
.61	.62	.57	.57	.57	.57	.57	.62	.61
.62	.61	.57	.57	.57	.57	.57	.61	.62
.61	.62	.57	.57	.57	.57	.57	.62	.61
.58	.67	.57	.57	.57	.57	.57	.67	.58
.51	.67	.67	.62	.61	.62	.67	.67	.51
.38	.51	.58	.61	.62	.61	.58	.51	.38

FIGURE 10.4b

harvests to provide forage. If we increasingly restrict the amount of forest that can be removed by imposing a policy constraint (equation 10.9), the partial removals (not shown) are reduced but remain distributed throughout the forest until we constrain harvests to about 38% of the original forest area. At that point, the center cell is left entirely unharvested, along with the four corner cells of the forest, and the other stands are still managed through fairly uniform partial removals. Efficient allocation of the resulting forage through equation (10.6) leaves no forage for a population in the center stand and a reduced amount of forage in the adjacent stand to the right (figure 10.5b). Such asymmetrical results are common in individual reaction–diffusion model solutions with areawide capacity limits, reflecting the potential for (perhaps many) somewhat random population arrangements in nature (see chapter 13).

As harvesting is further constrained to about 12% of the original forest area, the unharvested zone in the center of the forest enlarges, with partial removals concentrated around it, until a maximum size is reached that places

.30	.39	.45	.47	.48	.47	.45	.39	.30
.39	.52	.52	.48	.48	.48	.52	.52	.39
.45	.52	.44	.44	.44	.44	.44	.52	.45
.47	.48	.44	.44	.44	.44	.44	.48	.47
.48	.48	.44	.44	.44	.44	.44	.48	.48
.47	.48	.44	.44	.44	.44	.44	.48	.47
.45	.52	.44	.44	.44	.44	.44	.52	.45
.39	.52	.52	.48	.48	.48	.52	.52	.39
.30	.39	.45	.47	.48	.47	.45	.39	.30

FIGURE 10.5a

Optimal static equilibrium wildlife population distributions (in hundreds of breeding pairs) for a hypothetical species when forest nesting habitat provides no forage, with forest conversion to foraging habitat (a) unlimited and with forest conversion to foraging habitat limited to (b) about 38%, (c) about 12%, and (d) about 1% of the forest.

all of the population in the outer three rings of cells (figure 10.5c). With a further decrease in harvest area to about 1%, the total population declines, but the unpopulated area in the center of the forest changes little, as shown in figure 10.5d. The few partial removals are spread around the cells immediately adjacent to the unpopulated zone to provide forage for nesting pairs in that layer of stands.

If no harvests are allowed, the population drops to zero. Although the nest sites in the outer two rings of forest cells are within foraging distance of the surrounding nonforest, the usable habitat is too restricted, falling short of a critical extinction threshold (see chapter 6) imposed by the reaction–diffusion conditions in equation (10.3). Disregarding this species' foraging require-

.27	.36	.41	.43	.44	.43	.41	.36	.27
.36	.44	.44	.44	.44	.44	.44	.44	.36
.41	.44	.44	.44	.44	.44	.44	.44	.41
.43	.44	.44	.44	.44	.44	.44	.44	.43
.44	.44	.44	.44		.28	.44	.44	.44
.43	.44	.44	.44	.44	.44	.44	.44	.43
.41	.44	.44	.44	.44	.44	.44	.44	.41
.36	.44	.44	.44	.44	.44	.44	.44	.36
.27	.36	.41	.43	.44	.43	.41	.36	.27

FIGURE 10.5b

ments by managing as if it was an interior specialist (as in figure 10.3) would cause this population to perish.

Dynamic Scheduling

Next, we examine the full scheduling model as portrayed in equations (10.1)–(10.9) for our 9×9 grid of forest stands (still surrounded by 88 cells of nonforest) over a planning horizon of 10 time periods beyond the initial condition of mature forest at time $t = 0$. The parameter values for equations (10.3), (10.6), and (10.8) remain unchanged, and no policy constraints (equation 10.9) are imposed. The initial wildlife populations for equation (10.4) are set to about 65% of the values shown in figure 10.5a. Our management variables (X_{ik}), representing the portion of each stand assigned to a given schedule k, provide options for initially harvesting at the beginning of any of the 10

.13	.17	.20	.21	.21	.21	.20	.17	.13
.17	.23	.26	.26	.26	.26	.26	.23	.17
.19	.25	.27	.26	.16	.26	.27	.25	.19
.20	.25	.24				.24	.25	.20
.20	.24						.24	.20
.20	.24	.23				.23	.24	.20
.18	.24	.25	.23		.23	.25	.24	.18
.16	.21	.24	.24	.23	.24	.24	.21	.16
.12	.16	.18	.19	.19	.19	.18	.16	.12

FIGURE 10.5c

time periods (or not at all) and reharvesting at the start of any time period in which regrowth is 4 time periods old or older. Because we model harvests at the beginning of each period, the age of a stand chosen for harvest is reset to zero for that time period.

The nest site capacity coefficients for equation (10.5) are set to zero for forest harvest units at ages zero, one, and two periods and to 1 for ages three and older. The forage capacity coefficients for equation (10.7) are set to 1 at age zero, .75 at age one, and .5 at ages two and older. One unit of forage per cell is also still available in every time period from each of the 88 cells surrounding the forest.

It was necessary to iteratively apply the objective function (equations 10.1 and 10.2) eight times to maximize all 10 periods (populations were simultaneously binding for the last 3 periods). Table 10.1 shows the resulting total population over time. To characterize the results spatially, it also shows the forest age class distributions and populations for the center cell and one of the

Habitat Edge Effects 157

.03	.04	.04	.04	.04	.04	.04	.03	.03
.04	.05	.05	.05	.05	.05	.05	.04	.03
.04	.05	.06	.05		.05	.05	.05	.04
.04	.05	.05				05	.05	.04
.04	.05	.03					.05	.04
.04	.05	.05				.05	.05	.04
.04	.05	.06	.05		.05	.05	.05	.04
.04	.05	.05	.05	.05	.05	.05	.04	.03
.03	.04	.04	.04	.04	.04	.04	.03	.03

FIGURE 10.5d

corner cells. Like the population distributions from our static model (see figure 10.4), the largest cellular populations in this solution (not shown in table 10.1) occur in the cells close to (but not in) the corners.

As we would generally expect, the total population in table 10.1 steadily increases from initial conditions up to a new steady-state level (41.78 units in this case). Likewise, the corner cell population increases steadily and levels off, and the management schedules for that cell appear to produce a repetitive three-period cycle of age class distributions after an initial four-period conversion phase. Although the overall population across the forest has reached steady-state conditions, the habitat conditions (indicated by age class distributions) and population in the center cell (the most isolated from permanent forage) still appear to be in transition. From the other near-equilibrium conditions in this solution, it seems likely that even the center cell would settle into an equilibrium pattern given a longer planning horizon. It also seems clear from this solution that when habitat dynamics and management scheduling

TABLE 10.1

Periodic Total Population (in Hundreds of Breeding Pairs) and Corner and Center Cell Populations and Habitat Age Class Proportions as Scheduled with the Dynamic Model Using a Sequence of MAXMIN Optimizations

Period	Total Population	Corner Cell Age 0	1	2	3	4+	Population	Center Cell Age 0	1	2	3	4+	Population
0	23.62	—	—	—	—	1.00	.19	—	—	—	—	1.00	.29
1	27.56	.03	—	—	—	.97	.19	—	—	—	—	1.00	.42
2	32.61	.25	.03	—	—	.72	.20	.38	—	—	—	.62	.55
3	37.29	.49	.25	.03	—	.23	.23	.10	.38	—	—	.52	.52
4	37.60	—	.49	.25	.03	.23	.26	.07	.10	.38	—	.45	.45
5	39.36	.22	—	.49	.25	.04	.28	.05	.07	.10	.38	.40	.44
6	40.91	.13	.22	—	.49	.16	.30	—	.05	.07	.10	.78	.41
7	41.06	.33	.13	.22	—	.32	.32	.39	—	.05	.07	.49	.56
8	41.78	.22	.33	.13	.22	.10	.32	.09	.39	—	.05	.47	.52
9	41.78	.12	.22	.33	.13	.20	.33	—	.09	.39	—	.52	.42
10	41.78	.33	.12	.22	.33	—	.33	.36	—	.09	.39	.16	.54

are included, complex optimal strategies can arise from a fairly simple set of assumptions.

In our static models, a mechanical definition of edge may not be difficult to specify based on habitat contrast, such as a large difference in age between two adjacent stands. In dynamic models, however, the ages would be constantly changing, and arbitrary thresholds would have to be used to define when edge is present and when it is not. As the results in table 10.1 demonstrate, the actual effects of edge could be difficult to portray using a mechanical approach, and useful strategies such as partial removals could be overlooked entirely. Our model focuses on the underlying reasons why edge is important—multiple habitat needs, including the costs associated with their spatial separation—and provides a definition of edge that is useful in a dynamic scheduling context. In this chapter, we have represented both forest growth and wildlife population processes using identical time scales, but such an approach may often be unrealistic. When these processes function in significantly different time scales, a nested schedule approach similar to the one discussed in chapter 3 might be feasible.

Management Scale Effects

So far, we have considered only populations of wildlife species with breeding territories that are small compared to one forest stand land unit. That is, we have assumed that portions of each stand can be treated differently, with confidence that the nest sites will not be too finely divided. When our linear programming solution suggested a cellular population of .23 units, for example, this implied that at least 23 usable nest sites were retained.

For some species, a single stand (or other land management unit) may represent few or even less than one whole breeding territory. In such cases, mixed-integer programming with breeding pairs modeled as integer state variables may be more appropriate than linear programming to ensure that management schedules spatially locate adequate areas to support whole nest sites. In this experiment, we examine a static modeling case in which each forest stand represents a single breeding territory, and the S_i population variables are 0–1.

At this scale, our 9×9 stand forest is too small to support a persistent wildlife population with the net reproduction rate (r value) and dispersal parameter values we previously used for equation (10.3). Consequently, for this experiment we keep the dispersal parameter values unchanged but increase the r value to 3.0 to model a persistent wildlife population (see chapter 6). Other parameter values are again set so that unharvested forest stands provide 1 population unit of nest site capacity and .5 unit of forage. Harvested stands

1.0		1.0	1.0		1.0		1.0	1.0
1.0								1.0
	1.0			1.0				
1.0			1.0			1.0		1.0
								1.0
1.0		1.0			1.0			1.0
	1.0				1.0			
								1.0
1.0	1.0		1.0	1.0		1.0		1.0

FIGURE 10.6

A near-optimal static equilibrium wildlife population distribution (in breeding pairs) for a hypothetical species with whole-stand nest site requirements when forest nesting habitat provides .5 population unit of forage within each stand.

provide no nest site capacity and 1 unit of forage, like the surrounding non-forest cells.

The population distribution resulting from processing 150,000 nodes using a branch-and-bound solution algorithm is shown in figure 10.6. Although we cannot be certain that this solution is optimal, careful examination suggests that it is at least very good. Much of the population occurs in small clusters. This clustering probably results from reaction–diffusion effects from equation (10.3), and the small size of the clusters reflects the need to spread the population out for foraging purposes. Also, 20 of the 28 breeding pairs are located along the edge of the remaining forest close to forage. In this solution, the original forest is reduced to scattered remnant stands to improve conditions for a species that responds favorably to edge. This pattern of forest removal is very different from the unfragmented circular habitat patch that is typically most efficient for reaction–diffusion limited populations (Ludwig et al. 1979; Holmes et al. 1994) that respond negatively to habitat edge (as in figure 10.3).

In this examination of scale effects, we have not considered the case in which some other factor dictates a management unit smaller than the wildlife breeding territory. We anticipate that mixed-integer programming using an "aggregate emphasis" (or "coordinated allocation zone") formulation (Crim 1981; Johnson et al. 1986; Davis and Johnson 1987) may offer a suitable modeling method for such a species, but further work on this is needed.

To conclude, we note that we have modeled land units using fixed boundaries and treatment schedules that produce abrupt edges with no ecotones. When ecotonelike habitats can be artificially induced through management activities such as selective thinning or prescribed burning, they might be modeled in this same fashion. However, it is common for the abrupt edge produced by intensive tree removal to gradually become less abrupt over time as additional sunlight, wind, seed sources, and other influences change the microhabitat and species composition some distance into the remaining forest (Laurance and Yensen 1991). This process is not included in our model.

Despite the inability of our model to capture all the subtleties of a forest habitat and wildlife population, the results seem reasonable as a first-order approximation of effects. The model arises from a definition of edge that focuses on the basic wildlife habitat requirements that make edge important, whether positively or negatively so. This approach also allows a dynamic interpretation for purposes of scheduling management actions. In part III we switch our focus to the problem of using optimization models as tools for controlling the undesirable effects of pests and wildfires.

PART III

CONTROL MODELS

We have identified this part as "Control Models" because it actually makes an enormous difference to optimization modeling whether a population or other entity is to be maximized or minimized. In part II, populations were determined by whichever factor (e.g., carrying capacity, population growth, dispersal) is limiting. We are able to model this by maximizing the population (s) subject to the minimum limiting factor (f_i) as a function of management (x):

Maximize

s,

subject to

$s \leq$ the smallest $f_i(x)$,

which we can formulate as follows:

Maximize

$$s \tag{III.1}$$

subject to

$$s \leq f_i(x) \quad \forall i. \tag{III.2}$$

Incidentally, we can also solve optimization models that minimize s subject to the maximum limiting factor as follows:

Minimize

s,

subject to

$s \geq$ the largest $f_i(x)$,

which we can formulate as follows:

Minimize:

$$s \qquad\qquad\qquad (III.3)$$

subject to

$$s \geq f_i(x) \quad \forall i. \qquad\qquad\qquad (III.4)$$

In ecological control problems, we often want to minimize the population (or other entity) as in (III.3 and III.4), but the population will be determined by the smallest (most limiting) f_i, as in (III.1 and III.2). It is not reasonable to combine (III.3) and (III.2): The solution would incorrectly set all $s = 0$ with no management expediture. Incidentally, it is also not possible to maximize s subject to the largest f_i. In Hof and Bevers (1998), we addressed this problem with an integer formulation, which was very difficult to solve at even a small scale.

In part III we demonstrate two approaches to this dilemma in linear control models. In chapter 11, whose models are otherwise very similar to the reaction–diffusion models in part II, we solve the problem by combining the multiple limiting factors into a single equality constraint. With this approach, we address the problem of controlling an invading exotic pest. The management actions are defined as directly treating pest populations rather than the carrying capacity, as in part II. This causes the effects of management actions to be formulated as a part of the population growth and dispersal function, much like the captive-bred ferret releases in chapter 7, avoiding the problem of multiple potential limiting factors. This solution is tenable only when management can be associated directly with population (and not carrying capacity, as in host elimination). In this formulation, areas with innately limited carrying capacity or host presence (regardless of management) are accounted for in the dispersal coefficients.

In chapter 12, we address the problem of managing a potentially destructive wildfire. The intuitive approach for such a problem would be to minimize the fire's heat intensity or rate of spread, subject to whatever fuel (which can be managed), weather, or topography factors are limiting, but this leads to multiple less-than-or-equal-to limiting factors with a minimand (a minimization objective function). Also, in modeling the ignition of individual areas (e.g., cells), the limiting factor is not the fire source that could be the hottest, but rather the source that will ignite the given area first. Therefore our approach to this problem in chapter 12 is to focus on the timing of fire spread. This approach converts the problem to one of maximizing (delaying) ignition times subject to the minimum (first) ignition time across all potential ignition sources, a problem that can be formulated like (III.1) and (III.2). This approach also allows the fire to spread through the land area cells at different rates of speed, depending on fuel levels, which is not a feature of our formulations in part II. This model is very different from any of the others in the book and is based on an objective of delaying the ignition of certain priority areas (e.g., towns, homes) rather than fire suppression per se.

11

STRATEGIES FOR CONTROLLING EXOTIC PESTS

Recent literature documents the important and expanding problem of exotic pests invading forest ecosystems. As Haack and Byler (1993:34) summarize, "most native insects and pathogens reach a dynamic state of equilibrium with their hosts and natural enemies. However, this situation may not be true for . . . newly introduced exotic insects and pathogens." These authors also state, "Exotic insects and pathogens have dramatically altered forest ecosystem diversity, function, and productivity. More than 20 exotic fungal pathogens and 360 exotic insects now attack woody trees and shrubs in North America" (1993:33). These numbers have increased (and probably will continue to increase) as a result of increased human mobility, alterations of natural habitats and natural movement avenues, and increases in global transport of forest and other materials (Liebhold et al. 1992, 1995).

The invasion of exotic organisms involves three processes: arrival, establishment, and spread (Elton 1958; Dobson and May 1986). With regard to arrival, Liebhold et al. (1995:37) state, "Although some measures are currently being taken to prevent the transportation of exotic pests, relatively little additional effort in this area could substantially reduce the frequency of forest pest invasions." Liebhold et al. emphasize the importance of detecting new infestations and "eradicating introduced pests while their populations are small and localized" (1995:38).

This chapter investigates the possibility of using optimization methods to

This chapter was adapted from J. Hof, Optimizing spatial and dynamic population-based control strategies for invading forest pests, *Natural Resource Modeling* 11, no. 3 (1998): 197–216, with permission from the publisher, the Rocky Mountain Mathematics Consortium.

assist in developing a spatial management strategy for an invading pest once it has arrived and has at least minimally established itself. Obviously, preventing the pest's arrival or eradicating it immediately after its arrival is highly preferable, but it seems that some invading pests inevitably will manage to establish themselves, and the problem of determining an optimal management strategy over time and space is not trivial. Other authors (e.g., Reed and Errico 1987) suggest optimization formulations that account for expected pest-caused losses as they vary by timber classes, but these formulations do not directly account for pest growth and dispersal or determine spatial management strategies.

Liebhold et al. (1992, 1995) note that barrier zone projects, which focus on the expanding front of the infested area, can slow the rate of expansion of an established exotic. This chapter explores the possibility that more complex spatial and dynamic strategies should be considered and develops some simple modeling methods that might be useful in investigating those possibilities.

Formulation

Liebhold et al. (1995) suggest a reaction–diffusion formulation (following Skellam 1951) that is based on Fick's law of diffusion and an exponential growth model to predict the growth and dispersal of an invading organism. As in part II, we apply a model like this with discrete time periods and cells, treating each time period as a "time of infestation" and each cell as a "point of infestation." Again we capture the dispersal model with a matrix of coefficients, g_{nh}, that indicate the proportion of population that disperses from each cell, indexed with n, to each other cell, indexed with h, during one time period (and using cell centroids to calculate distances). The diagonal of this coefficient matrix, where $n = h$, indicates the proportion of the population that remains in each cell from one time period to the next. We thus model the pest population in each cell h in each time period with a difference equation like the ones we used in part II:

$$S_{ht} = \sum_n g_{nh}\left[(1 + r_n)\, S_{n(t-1)}\right] \quad t = 1, \ldots, T \tag{11.1}$$

$$\forall h$$

$$S_{n0} = N_n \qquad\qquad \forall n,$$

where
   h and n both index cells,
   t indexes time periods ($t = 0$ for initial conditions),

r_n = the r value for cell n,
S_{ht} = the pest population in cell h in time period t,
$S_{n(t-1)}$ = the pest population in cell n in time period $t - 1$,
N_n = the initial pest population in cell n.

Liebhold et al. (1995:6) note,

The above model assumes that r [r value] and D [diffusivity or diffusion coefficient] are constant through time and space during the period
of range expansion of the invading organism, an assumption that does
not intuitively seem likely (e.g., spatial variation in the habitat may pro-
foundly affect birth/death functions as well as dispersal rates). Never-
theless there has generally been good congruence between predictions
of this model and observed rates of spread of most exotic organisms
(Levin 1989, Andow et al. 1990). For example, Long (1979) found that
the larch casebearer, *Coleophora laricella,* has been spreading in the
northern Rocky Mountains at a constant radial rate, as predicted by
Skellam's (1951) model.

The g_{nh} coefficients in (11.1) can be varied to reflect spatial variation across
the landscape (and are varied in the following case example) but are assumed
constant across all time periods. This means that no limits to host capacity are
included in (11.1), just as they are not included in the Liebhold et al. (1995)
model. Varley et al. (1973) discuss the inclusion of host or carrying capacity
limits in this kind of model, but such inclusion in the optimization model
would complicate solution by necessitating integer variables (see Hof and
Bevers 1998). We therefore ignore host capacity limits here. This implies that
pest population numbers probably will become meaningless if the population
gets out of control except to indicate the out-of-control condition. We discuss
spatial variation of the g_{nh} coefficients further in the case example.

If it is possible to reduce pest population through spatially specific management actions, then (11.1) could be augmented as

$$S_{ht} + \alpha_h Y_{ht} = \sum_n g_{nh} \left[(1 + r_n)\, S_{n(t-1)} \right] \quad t = 1, \ldots, T \qquad \forall h \quad (11.2)$$

$$S_{n0} = N_n \qquad \forall n,$$

where
Y_{ht} = the amount of management effort in cell h and time period t,
α_h = the pest population exterminated per unit of Y_{ht} in cell h.

This formulation (11.2) forms the basis of the optimization model suggested here. Any number of economic or biological objective functions could be used, but for simplicity let us assume that we want to minimize the total exotic pest population over T time periods. Let us also assume that our management resource limitation can be captured by a simple constraint that specifies that we are able to apply only K_t units of management in each time period t. The problem, then, is to optimally locate the management-induced pest mortality over time, given the initial pest infestation location and subsequent growth and dispersal (as it is affected by the management actions).

The basic optimization model is then as follows:

Minimize

$$\sum_{t=1}^{T}\sum_{h} S_{ht},\tag{11.3}$$

subject to

$$S_{n0} = N_n \qquad\qquad\qquad \forall n,\tag{11.4}$$

$$S_{ht} + \alpha_h Y_{ht} \geq \sum_{n} g_{nh}\left[(1+r_n)\,S_{n(t-1)}\right] \quad t = 1,\dots,T, \quad \forall h,\tag{11.5}$$

$$\sum_{h} Y_{ht} \leq K_t \qquad\qquad\qquad t = 1,\dots,T.\tag{11.6}$$

Equation (11.5) is written as an inequality in the model to aid in avoiding numerical problems in model solution. This inequality allows the possibility of unnecessary management actions if (11.6) is not binding. We can eliminate this possibility by adding

$$\alpha_h Y_{ht} \leq \sum_{n} g_{nh}\left[(1+r_n)\,S_{n(t-1)}\right] \quad t = 1,\dots,T \quad \forall h,$$

or with "roll-over runs" that eliminate slack in (11.5). Note that (11.6) is non-binding in a given time period only if the pest population is zero in that time period. In the case example, management layouts are reported only for time periods where pest population is nonzero, so this is not a problem.

This model assumes that management can be meaningfully related to the exterminated pest population, Y_{ht}. For situations in which management options must be related to host elimination (e.g., sanitation cutting of a host tree species), the reader is referred to the approaches in Hof and Bevers (1998). The assumption that management effort is directly related (not necessarily linearly) to pest mortality actually simplifies the problem significantly. If management effort can be directly related to pest mortality but with diminishing returns, this can be included in the preceding optimization model without violating convexity requirements through piecewise approximation. This would be formulated by replacing (11.5) and (11.6) with

$$S_{ht} + \sum_{j=1}^{J} \alpha_{jh} Y_{jht} \geq \sum_{n} g_{nh} \left[(1 + r_n) S_{n(t-1)} \right] \quad t = 1, \ldots, T, \quad (11.7)$$
$$\forall h,$$

$$Y_{jht} \leq Q_{jht} \qquad\qquad\qquad\qquad t = 1, \ldots, T \quad (11.8)$$
$$\forall h$$
$$\forall j,$$

$$\sum_{j} \sum_{h} Y_{jht} \leq K_t \qquad\qquad\qquad t = 1, \ldots, T, \quad (11.9)$$

where

j indexes the segments of Y_{jhf} for piecewise approximation of the non-linear relationship between management effort and pest mortality,

Y_{jhf} = the amount of management effort in cell h, segment j, and time period t,

α_{jh} = the pest population reduction that results from each unit of management effort in segment j and cell h,

Q_{jht} = the size of the jth segment for cell h and time period t.

Diminishing returns (diminishing management effectiveness) implies that

$$\alpha_{1h} > \alpha_{2h} > \ldots > \alpha_{Jh} \quad \forall h$$

which ensures convexity of the model.

Case Example

We present a simple example of the modeling approach suggested. The purposes of this example are to demonstrate the potential application of the model formulation and to show that simple spatial management strategies such as the barrier zone approach to slowing an organism's invasion may not always be optimal. For these purposes, we constructed a stylized model with no empirical connection to any particular exotic pest or geographic location.

We began with a 30 × 30 grid of 900 cells to define the spatial planning area. Six time periods ($t = 0, \ldots, 5$) were adequate for the desired demonstrations, and we experimented with a variety of initial infestation locations and K_t levels. Throughout this chapter, we assume an initial infestation of 2 population units in each of 25 cells, arranged in a square set of cells (5 × 5, with 50 population units total). We applied an r value of 9, such that the untreated pest grew tenfold in population ($1 + r = 10$) during each time period. We used the simpler formulation in equations (11.3)–(11.6), which reflects constant effectiveness of management effort in pest extermination. This simple approach makes it possible to eradicate the pest in some of the solutions presented here. This is acceptable for a stylized example but may be overly optimistic for many real-world problems, where the formulation in (11.7)–(11.9), which accounts for diminishing returns in management effectiveness, might be more realistic. For simplicity, we assumed that the α_h is 1 for all cells, so management effort is measured in terms of pest population exterminated, and management effectiveness is uniform across all cells. We initially assumed that for each cell, h, the g_{nh} for the 8 immediately surrounding cells is .05625, the g_{nh} for the 16 cells that are one removed is .015625, and g_{nh} is zero for all cells that are more than one removed. When $n = h$, the g_{nh} coefficient is .3. The g_{nh} coefficients for cells outside the planning area reflect an absorbing boundary assumption that no pests immigrate from outside the planning area and pests that disperse outside the planning area die immediately (or are ignored). This assumption creates opportunities that can be used to advantage in the optimization process. In any real-world application, the g_{nh} could be based on calculations from a model like that in Liebhold et al. (1995) and could also be adjusted to account for partial landscape barriers to dispersal, prevailing winds, special dispersal routes (both natural and human-made) such as rivers or trade routes, and any number of other spatial factors.

As hypothesized, the model example represents a pest with strong reproduction and dispersal capability. Note that the initial ($t = 0$) time period is used only to set the initial pest infestation, and management is allowed only in time periods 1–5. The population in each time period reflects the growth and dispersal from the previous time period and the pest mortality from management applied in that time period. The dimensions of the case example create a linear program with approximately 6300 constraints, 10,800 choice variables, and 131,600 nonzero coefficients.

Results

It should be noted that with all the solutions presented, many alternative optima exist. Also note that in all figures, the population shown in each time period reflects the impacts of the management actions in that period. Figure 11.1 shows

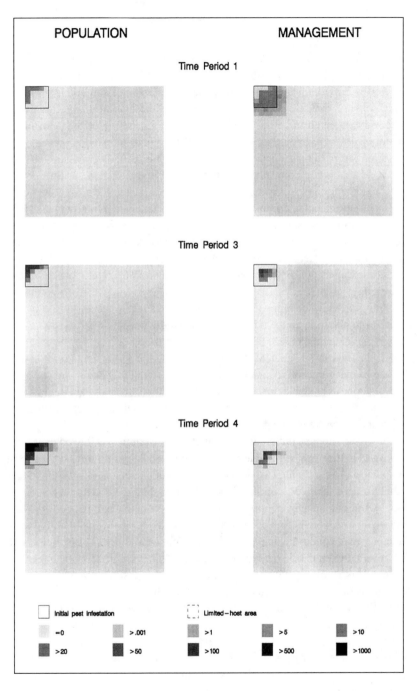

FIGURE 11.1

Model solution for selected time periods, with the initial infestation as shown and $K_t = 350$ for all t.

the model solution, with the initial infestation placed in the upper left-hand corner of the planning area and with K_t set at 350 population units in each time period 1–5. With this level of management available, the solution looks much like some of the case histories documented in the literature (e.g., the gypsy moth documented in Liebhold et al. 1992, 1995). In time period 1, management effort focuses on the border zone of the initial infestation area and the interior part of that area near the border zone. This leaves a population in time period 1 that is contained near the corner of the planning area. By time period 3, the population is still contained in this same area, but the size of the population is getting out of control (table 11.1). The management strategy is strictly focused on the border zone at this point, with eradication being impossible. By time period 4, the population is beginning to expand away from the edge of the planning area, and the management effort is backing away as the border zone moves out. By this time, the pest population is totally out of control (table 11.1).

In figure 11.2, the initial infestation is moved south to the north- south middle of the planning area but is still up against the west (left-hand) edge. Also, the K_t is increased to 400 population units in time periods 1–5. As in figure 11.1, the initial management strategy in time period 1 is to contain the spread in the border zone and to attack as much of the initial infestation area near the border zone as possible. In time periods 1–3 (1 and 2 are shown in figure 11.2), the management effort is able to contain the pest on the edge of the planning area and nearly keeps the pest population constant (table 11.1). In time period 4, however, the pest population begins to escape control, followed by a large increase in time period 5 (table 11.1). In figure 11.2, by time period 5, the pest population has split into two populations, and management effort is no longer containing it. This solution shows how the modeling method might be used to show times and places of opportunity. If some extra management effort could be made available before time period 4 along the edge of the planning area, the outbreak conditions in time period 5 might be avoided. This solution also shows that even if a pest population appears to be stabilized by management effort, this stability may be temporary until the pest gains enough spatial advantage to dissipate the management's effectiveness. Put another way, the effectiveness of a border zone strategy may erode over time if the pest is able to expand the size of the border zone. The patterns in figure 11.2 also begin to show the importance of landscape features in optimizing pest management. The west (left) edge of the planning area serves as a landscape feature that the management strategy takes advantage of in trying to contain the pest. Interior landscape features are examined in figures 11.5 and 11.6.

In figure 11.3, the initial infestation is moved to the center of the planning area to eliminate the effect of the planning area edge. Also, the K_t is set at 450

TABLE 11.1

Pest Populations Associated with Figures 11.1—11.6

Time Period	Figure 11.1	Figure 11.2	Figure 11.3	Figure 11.4	Figure 11.5	Figure 11.6
0	50.0	50.0	50.0	50.0	50.0	50.0
1	82.4	64.4	50.0	50.0	50.0	50.0
2	212.8	64.7	50.0	50.0	50.0	50.0
3	1032.9[a]	67.1	50.0	50.0	28.0	49.9
4	6443.0[a]	84.4	50.0	50.0	0.0	43.7
5	45,835.9[a]	209.3	50.0	44.2	0.0	0.0
6	N/A	N/A	N/A	0	N/A	N/A
Total	53,657.0[a]	539.9	300	294.2	178.2	243.6

[a]Number is meaningless other than to indicate that the pest is out of control.

population units in time periods 1–5. This level of management is just able to keep the pest population stable at 50 units without the influence of any spatial features ($1 + r = 10$, so a pest population of 50 grows to 500 in each time period, and the 450 units of management-induced population reduction are just enough to compensate). In figure 11.3, the initial management strategy is to try and cover as much of the initial infestation area and the border zone surrounding it as possible. The pest population is indeed stable (table 11.1), but by time period 4 the pest is splitting into scattered populations, and spatial containment is lost (figure 11.3). The only hope for eradication in this type of scenario (other than increasing the allowed management effort) would be to "move" the pest to one (or more) of the planning area edges by concentrating available extermination effort in some areas and allowing the pest to spread into other areas. Looking at figure 11.3, this might be possible with many time periods but would be difficult because of the size of the management area relative to the allowed management effort.

In figure 11.4, the planning area is reduced to 625 cells (25×25) to see if the model can pursue such a strategy. For the graphic display, individual cell size is increased accordingly. The initial infestation is still the same size and located in the middle of the planning area, and K_t is left at 450 population units per time period. The initial strategy in time period 1 is much like that in figure 11.3, but in time periods 3 and 4 a clear (and successful) strategy emerges. In time period 3, the placement of management effort is beginning to create two populations: a smaller one in the southeast part of the planning area and a larger one that is being "moved" toward the north and west edges of the

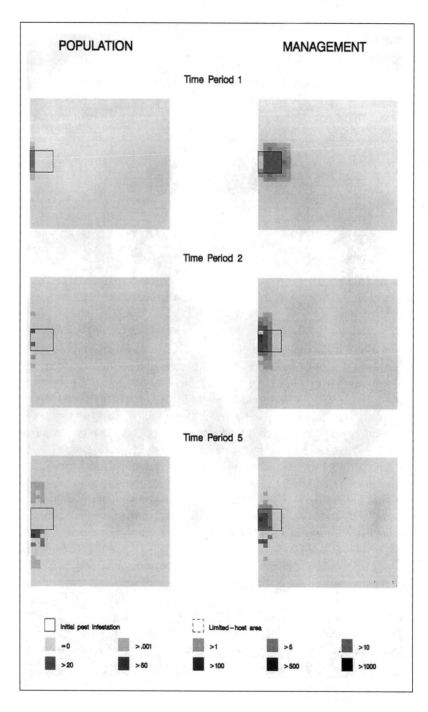

FIGURE 11.2

Model solution for selected time periods, with the initial infestation
as shown and $K_t = 400$ for all t.

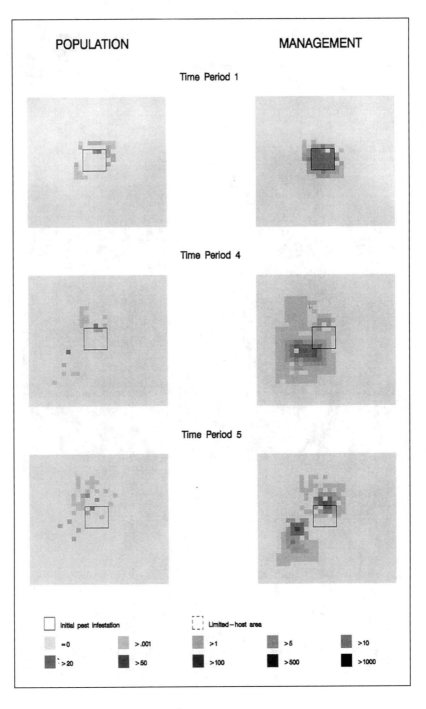

FIGURE 11.3

Model solution for selected time periods, with the initial infestation
as shown and $K_t = 450$ for all t.

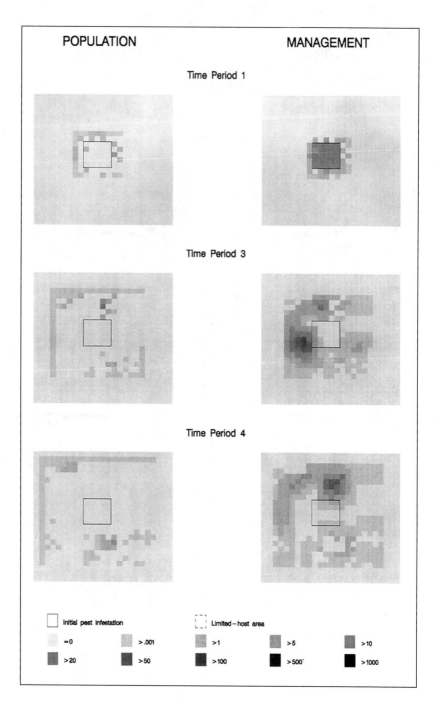

FIGURE 11.4

Model solution for selected time periods, with the initial infestation as shown and $K_t = 450$ for all t, with the planning area reduced to 625 cells (25 × 25).

planning area. By time period 4, the edges of the planning area are helping to reduce this larger pest population. By time period 5 (not shown in figure 11.4), the population in the northwest is well contained in the corner, and management effort is more concentrated on the population in the southeast. At this point, the population is coming under control (table 11.1). We had to add one more time period to the model to allow eradication, which takes place in time period 6 as management effort is able to finish off both populations (not surprisingly, because the population in time period 5 was 44.179; $10 \times 44.179 = 441.79$, which is less than the 450 allowed management units). It was not possible in time periods 2–5 to "move" the entire population into the corner of the planning area, but enough was "moved" to create the eradication situation in time period 6. Such a spatial management strategy is subtle and might not be considered without an optimization analysis. Figures 11.1–11.4 use the edges of the planning area to control the pest. In figures 11.5 and 11.6, we consider some interior landscape features that could be used to advantage.

As noted earlier, because the g_{nh} coefficients are defined pairwise, they can vary to reflect any number of spatial characteristics, including prevailing winds, special movement corridors, and movement obstacles. For example, suppose some portion of the planning area has much less host capacity than the rest of the area, causing high pest mortality for individuals that attempt to disperse into those areas. We will assume that the pest's success in finding hosts is a function of the host density in any given cell and that the proportion of pests that do find a host in an area of lower host density then behave as any other pests do in terms of reproduction and dispersal. To account for such a scenario in equation (11.5), we must modify the g_{nh} coefficients for each destination cell ($h \in \Omega$, where Ω is the set of cell indexes that have limited host capacity). We assume that these coefficients are one-tenth of the other g_{nh} coefficients, indicating that any time any of the pest population disperses into the limited-host area, 90% of that population perishes and 10% finds hosts and lives on, behaving as the other pest population does.

In figure 11.5, the planning area is returned to 900 cells, and the K_t is again set at 450 population units in each time period. The initial infestation area is again in the middle of the planning area, but now a limited-host area is included for all cells east of the twentieth column (two cells removed from the initial infestation area). The management strategy in time periods 1–3 (figure 11.5) is clearly to focus efforts outside the limited-host area and to take advantage of its inhospitable conditions for the pest. As in figure 11.4, not all the pest outside the limited-host area can be eliminated immediately, but enough is eliminated to allow eradication in time period 4 (table 11.1). The pest population that survives outside the limited-host area in time period 3 (figure 11.5) is the last to be exterminated (in time period 4, table 11.1).

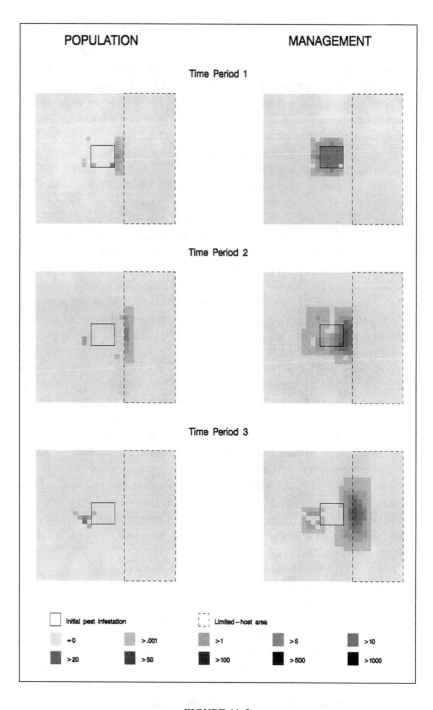

FIGURE 11.5

Model solution for selected time periods with the initial infestation as shown and $K_t = 450$ for all t, with the 11×30 cell limited-host area as shown.

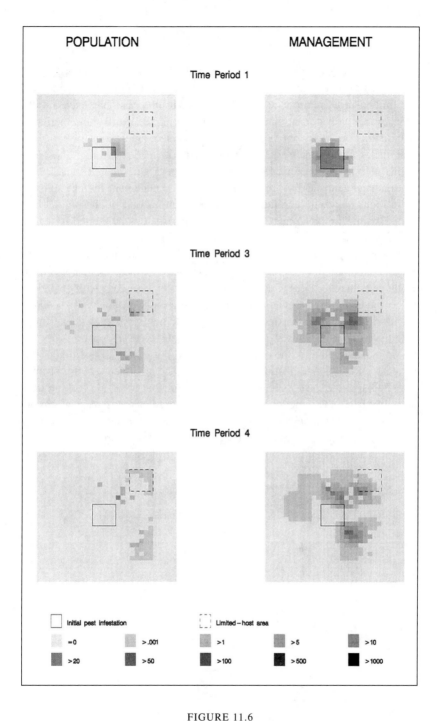

FIGURE 11.6
Model solution for selected time periods, with the initial infestation as shown
and $K_t = 450$ for all t, with the 5×5 cell limited-host area as shown.

In figure 11.6, the limited-host area is reduced to a 5×5 area northeast of the initial infestation area. As before, the strategy is to move the pest toward the limited-host area to take advantage of it. In this case, however, a much more spatially dispersed management strategy is needed. By time period 3, two distinct pest populations and several dispersed populations have formed. The management strategy is cohesively focused outside the limited-host area at this point. By time period 4, enough of the pest is in the inhospitable area that management starts to "mop it up," but the management strategy splits its main focus between the remaining two populations outside the limited-host area. By time period 4, much of the pest has perished in the limited-host area, setting up eradication in time period 5 (table 11.1).

Figures 11.5 and 11.6 involve very simple landscapes but demonstrate the mechanics and ability of the model to use interior landscape features to its advantage. In real-world problems, a very complex landscape with gradations of host availability and many other features might be included. By accounting for these features with the pairwise g_{nh} coefficients, complex management strategies might be devised to take advantage of the complex landscapes.

As indicated before, the case example does not reflect any particular forest pest, and no generalizations are intended. The approach requires specific knowledge and measurements of pest behavior that are often lacking or at least imperfect. The model must necessarily predict processes that are not completely understood and are largely random, so (like the rest of our models) it would best be applied in an adaptive management process. In chapter 12, we turn our attention to the problem of controlling wildfires.

12

STRATEGIES FOR CONTROLLING WILDFIRE

The use of mathematical models for managing fires has a rich and varied history in North America. Martell (1982) traces the application of operations research methods in forest fire studies back to the early 1960s. The earliest potential applications of operations research techniques to wildland fire management were mentioned by Shephard and Jewell (1961). Follow-up work by Parks and Jewell (1962) generated considerable interest by examining the use of differential equations and calculus to identify the optimal suppression force for a forest fire. Swersey (1963) and McMasters (1966) extended Parks and Jewell's work by focusing on the optimal mix of different suppression units and the effects of labor constraints on resource allocation rules.

Growing familiarity with optimization techniques spawned additional fire management applications, notably analyses of detection options (Kourtz and O'Regan 1971), airtanker retardant delivery systems (Simard 1979; Greulich and O'Regan 1982), and fuel management (Omi et al. 1981). Fire suppression has continued to receive attention through the use of optimal control theory (Parlar and Vickson 1982), nonlinear programming (Aneja and Parlar 1984), and catastrophe theory (Hesselyn et al. 1998). In addition to optimization, simulation modeling has provided useful insights for evaluating management alternatives, especially in an uncertain decision environment (Ramachandran 1988; Fried and Gilless 1988; Mees and Strauss 1992; Mees et al. 1993; Gilless and Fried 1999). Other simulation work oriented toward allocating management

This chapter was adapted from J. Hof, P. N. Omi, M. Bevers, and R. D. Laven, A timing-oriented approach to spatial allocation of fire management effort, *Forest Science* 46, no. 3 (2000):442–451, with permission from the publisher, The Society of American Foresters.

resources in fire containment efforts includes Mees (1985), Anderson (1989), and Fried and Fried (1996). A number of expert systems have also been developed for wildfire containment applications, such as those of Kourtz (1987), Saveland et al. (1988), Fried and Gilless (1989), Stock et al. (1996), and Hirsch et al. (1998).

Recent interest in fire and fuel management is particularly motivated by the high fuel levels in many forested areas throughout the country, especially in ecosystems with short fire return intervals and areas where fire has been excluded during the 20th century (Arno and Brown 1991; Covington and Moore 1994). These fuel conditions create at least two management problems: the long-term reduction of fuel loads and the protection of lives and property from the fires that occur in the interim. These problems present critical challenges for fire managers, suggesting the need for new and spatially explicit management science methods expanded to landscape scales.

Traditionally, fire managers have focused on fire suppression strategies that emphasize direct control of the fire or containment of its perimeter within predetermined or natural barriers. When confronted with fires that exceeded control or containment capabilities of available suppression resources, the fallback position has called for protecting valued resources (typically human developments). This approach is also relevant when it is decided to let a fire burn to restore natural fire processes or as a part of fuel reduction efforts. Recent policy changes (Zimmerman and Bunnell 1998) call for an expansion of strategies for managing fires, especially at the landscape scale. One option between letting a fire burn unhindered and attempting suppression is to slow its spread across the landscape in the direction of valued resources.

In this chapter, which is oriented toward affecting the timing of fire spread, we assume that there are distinct areas of concern (e.g., towns, summer homes, campgrounds, ski areas) and that a fire management objective is to delay ignition of those protection areas as long as possible. The advantages of such a delay include maximizing the chances that other suppression efforts or independent factors such as weather changes might cause the fire to subside before the protection areas are affected and maximizing the time available to build fire lines around the protection areas, modify fuels to reduce a fire's severity near the protection areas, or evacuate the protection areas. To explore the use of timing variable models for managing ignition delays in protected areas, we experiment with a model of the fire management effort (Simard 1976) expended during a fire event. Retardant drops, for example, are commonly used to delay fire spread when the scale of the fire permits (Brown and Davis 1973). For larger incidents, actions such as backburning (possibly with aerial ignition) could be applied. For very hot fires, the management effort might involve a combination of tactics. An extension of this model to address longer-term actions such as fuel reduction programs also is presented.

Formulation

We begin by defining a grid of square land cells to capture the spatial allocation of management effort. As noted earlier, we limit the notion of fire management effort to describe activities (such as retardant application) aimed at slowing the spatial spread of a fire through each cell after ignition. We then define the management choice variables as the proportion of each cell that receives the given management effort. Alternatively, we could define the choice variables by the magnitude of delay effort applied to the entire cell. We assume that allocation of such effort to a given cell has the effect of slowing the fire front's movement through that cell, thus aiding in maximizing the objective of delaying ignition of the protection areas.

Given a wildfire ignition together with the wind, moisture, fuel, topography, and other conditions, fire prediction models (e.g., Andrews 1986; Finney 1998) could be used to predict the fire spread direction and the burn time of each cell on the basis of the management effort applied in that cell (and the effect it has on the fuel available to burn). For simplicity, we assume that the accelerating effect of larger available fuel amounts takes place within the scale of each cell and that there is no cell-to-cell spread rate interaction. In other words, the spread time through a given cell is assumed to be determined by its own available fuel, not by the available fuel or fire intensity of adjacent cells. For example, if a cell is ignited by a high-intensity fire but has a small amount of available fuel, we assume that the fuel level brings the fire back to the same rate of spread that would have occurred with a lower intensity ignition. This assumption probably is scale dependent and is more tenable for large cells (such as the 100-ha cells we use in our case example) than for small ones. Therefore the model is applicable only when the scale of the problem is large enough to not violate this assumption significantly but is still small enough to allow some kind of fire management effort. We also ignore the possibility of spot fires igniting ahead of the fire front.

For our timing-based approach, we define variables that track the time when the fire front enters each cell (ignition time) and when it exits each cell. Because any given fire might enter (exit) each cell from (to) different directions, some geometric approximation is necessary. We envision the burning of each cell as a fire front the width of the cell, igniting it at one side and exiting on the opposite side, even though cells might actually burn from corner to corner, from one side to an adjacent side, and so forth.

For ease of presentation, we assume that a single cell represents the fire origin. We initially assume a single protection area one cell in size, but this assumption is relaxed later. A model to delay the protection cell's ignition time as much as possible would be as follows:

Maximize

$$T_{mn}^{\circ},\tag{12.1}$$

subject to

$$T_{ab}^{\circ} = 0,\tag{12.2}$$

$$T_{ij}^{\circ} \le T_{hk}' \qquad \forall (h, k) \in \Omega_{ij}\tag{12.3}$$
$$\forall i, j,$$

$$T_{ij} - T_{ij}^{\circ} = f_{ij}(F_{ij}) \qquad \forall i, j,\tag{12.4}$$

$$F_{ij} = \overline{F}_{ij} - \left(\gamma_{ij}\overline{F}_{ij}\right) X_{ij} \quad \forall i, j,\tag{12.5}$$

$$0 \le X_{ij} \le 1 \qquad \forall i, j,\tag{12.6}$$

$$\sum_{i}\sum_{j} X_{ij} \le \overline{X}.\tag{12.7}$$

Indexes

i indexes cell rows, as does h,
j indexes cell columns, as does k.

Variables

T_{ij}° = the time at which the fire front ignites cell ij,
T_{ij}' = the time at which the fire front leaves cell ij,
T_{hk}' = the time at which the fire front leaves cell hk (and potentially
   ignites cell ij),
F_{ij} = the fuel available for combustion in cell ij after treatment,
X_{ij} = the proportion of cell ij allocated for treatment.

Parameters

a, b = the row and column of the fire origin cell,
m, n = the row and column of the protected area cell,
Ω_{ij} = the set of row and column indexes for cells that can *potentially*
   ignite cell ij (typically some subset of the cells adjacent to cell ij,
   determined primarily by a combination of wind conditions and
   topography),
f_{ij} = an empirical function that relates available fuel in cell ij to the
   duration of time between entry and exit of the fire front,

$\overline{F}_{ij}$ = the available fuel in cell ij before treatment,
γ_{ij} = the proportion of fuel that can be made unavailable for combustion with treatment on any area within cell ij (up to the entire cell),
$\overline{X}$ = the total number of cells that can be treated.

Equation (12.1) is the objective function and maximizes the time at which the protected area is subject to ignition by the fire's front. Equation (12.2) sets the ignition time of the origin cell to zero. Equation (12.3) relates the ignition time of each cell to the times at which the fire front leaves the cells that can potentially ignite it. The multiple inequalities in (12.3) cause each cell to be ignited by the first potentially igniting cell from which the fire front departs. Equation (12.4) relates the time it takes the fire front to move through each cell to the available fuel in that cell, given existing weather and topography. In practice, the spread rate function f would be estimated with state-of-the-art fire prediction models or empirical data. For exploratory purposes, we assume a nonlinear function relating spread rate to available fuel. Linear approximation of this f function is discussed in the case example. Equation (12.5) relates the available fuel in each cell to the treatment applied to that cell. We formulate the relationship between fire front duration and treatment levels in (12.4) and (12.5) through the effect of treatment on available fuel, but (12.4) and (12.5) could be replaced with a function that directly relates treatment to fire front duration. Equation (12.6) limits the treatment variables to be between zero (no treatment) and one (treatment of the entire cell). Equation (12.7) reflects the typical situation in which fire management resources are limited, so only a certain number of cells can be treated. This scarcity of resources creates the optimization problem.

We note that this model does not account for the logistics of applying treatment. For example, if it is impossible to deliver treatment to a given cell hk before the time indicated in the solution value for T_{hk}^0, then the solution would not be implementable. In the case example, we force $X_{ab} = 0$ to reflect the fact that this cell has ignited at time zero and treatment is not feasible. A larger initial exclusion period could easily be created by setting more cells' treatment variables to zero (to allow time for the initial attack to begin). Beyond this, however, providing for delivery logistics would be difficult in the model as formulated.

Case Example

To create a dramatic demonstration of the model, we assume a very hot fire and a large-scale problem area, defined with 400 square cells 100 ha in size (each cell is 1000 m $\times$ 1000 m) and an initial fuel load $(\bar{F}_{ij})$ of 6000 metric tons in each cell. We assume that the fire starts in a single cell and, given wind conditions and topography, can spread (from cell to cell) either south, east, or diagonally to the southeast. Thus $\Omega_{ij} = [(i-1,j), (i,j-1), (i-1,j-1)]$. This assumption represents a fire burning under prolonged extreme fire danger conditions driving the fire in a predominant (i.e., downwind) direction. More complicated directional patterns (e.g., following topographic variations) could be included directly in the model formulation if suitable estimates were available. We locate the origin cell at the upper left-hand corner of the grid ($a = b = 1$) because the grid then captures the possible expansion area of the fire for 20 cells (in all possible directions). The model grid typically would be defined taking the fire origin and expected movement into account. As noted earlier, we set $X_{1,1}$ to zero to reflect an assumption that there is insufficient time for treatment to be feasible in the origin cell. We assume initially that there is only one protection area cell where $i = j = 20$. The definition of the grid should reflect protection areas and the fire origin. Additional protection area cells are added later.

In our model, treatment creates a 50% reduction in available fuel, so $\gamma_{ij} = .5$ for all cells. Equation (12.4) indicates that the burn time of each cell is related to the fuel available in that cell after treatment. As previously noted, these relationships would be either empirically based or estimated with fire behavior simulation models (taking the fire's environment, such as fuel type and moisture, weather, and topography, into account). The f_{ij} functions in equation (12.4) would be estimated on this basis as inputs to our model. These functions typically would be asymptotic to both burn duration and available fuel axes, and they would be convex to the origin. If we define $D_{ij} = T'_{ij} - T^{\circ}_{ij}$ as the duration variable, a simple function that has these properties is

$$D_{ij} = \frac{\alpha_{ij}}{F_{ij}}, \tag{12.8}$$

where α_{ij} is an empirical parameter. For our model, we approximate this function (for each cell) between the initial fuel load $\bar{F}_{ij}$ and F^*_{ij}, defined as the available fuel if the cell is fully treated ($X_{ij} = 1$). If D^*_{ij} is the duration indicated by (12.8) for F^*_{ij}, then (12.8) (and (12.4)) can be approximated between $\bar{F}_{ij}$ and F^*_{ij} with the linear function

$$D_{ij} = T'_{ij} - T^{\circ}_{ij} = D^*_{ij} - m_{ij}(F_{ij} - F^*_{ij}) = \left(D^*_{ij} + m_{ij}F^*_{ij}\right) - m_{ij}F_{ij}, \tag{12.9}$$

where

$$F_{ij}^* = (1 - \gamma_{ij})\overline{F}_{ij} \quad \text{by (12.5)},$$

$$m_{ij} = -\frac{D_{ij}^* - \overline{D}_{ij}}{F_{ij}^* - \overline{F}_{ij}} = \frac{D_{ij}^* - \overline{D}_{ij}}{\gamma_{ij}^* \overline{F}_{ij}}.$$

The m_{ij} is the absolute value of the slope of the approximation. We set $\alpha_{ij} =$ 250,000 for all cells (which implies a fire's frontal velocity that approaches 1.5 km/hr when $F_{ij} = 6000$ metric tons) and then used (12.8) to calculate the D_{ij}^* and $\overline{D}_{ij}$ for each cell based on its $\overline{F}_{ij}$. For example, if $\overline{F}_{ij} = 6000$, then $F_{ij}^* = 3000$, and (12.8) would predict a duration for an untreated cell ($\overline{D}_{ij}$) of $41\tfrac{2}{3}$ minutes and a duration for a fully treated cell (D_{ij}^*) of $83\tfrac{1}{3}$ minutes. The duration response for fuels (and treatment levels) between these endpoints is linear-interpolated by (12.9) in the model. The error from the linear approximation between $\overline{F}_{ij}$ and F_{ij}^* is not large and probably is consistent with the precision we might expect in estimating an f function.

The parameter set assumed implies a very hot fire, one that in reality might be very difficult (or impossible) to manage, but the intent of the example is to provide a dramatic presentation, not to reflect any particular real-world situation. By setting $\overline{F}_{ij}$, γ_{ij}, and α_{ij} to the same values for all cells and by including the same fire spread directions for all cells, we have created a homogeneous landscape. We did this to provide an interpretable demonstration, but such assumptions would not be necessary in an application.

Results

In all figures that follow, the top half (a) shows the cells that are binding in terms of the time of ignition in cell (20, 20). These cells are identified by a positive "dual" or "shadow" price associated with constraint (12.4). The bottom half (b) of each figure shows the treatment solution values (for the X variables). Figure 12.1 presents the optimal solution with a total of 100 cells allowed to be treated ($\overline{X} = 100$). The time of ignition for cell (20, 20) is 1436 minutes.

When faced with a problem such as this example, a fire incident commander might be inclined to focus the attack around either the origin cell or the protection cell. The solution shown in figure 12.1 is a combination of both strategies, along with some effort allocated in between, creating something of an hourglass-shaped layout of effort. The width of the hourglass appears to be associated with the width of the binding burn path in figure 12.1a. It should be emphasized that cells outside this binding path still burn in the model, but they are not binding in the solution that maximizes $T_{20,20}^\circ$. Also, this binding

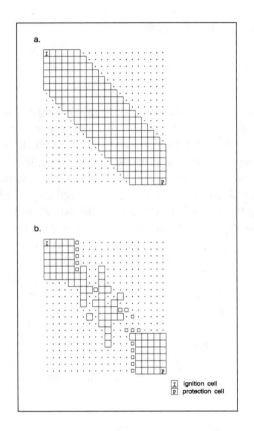

FIGURE 12.1

Binding burn path (a) and treatment areas (treatment amounts indicated by size of square for each cell) in optimal solution (b), with $\overline{X} = 100$ and one protection cell (20, 20).

path reflects (and is partially determined by) the treatment variable (X) solution values.

In figure 12.2, $\overline{X}$ is set to 18 cells (the convenience of this number will become apparent). The "hourglass" strategy collapses to two separate treatment areas around the origin and the protection cells. The time of ignition at cell (20, 20) is 1025 minutes, which is almost seven hours sooner than in the solution shown in figure 12.1. And the binding burn path narrows in response to the reduced amount of treatment allowed. An even narrower burn path than that in figure 12.2 occurs in optimal solutions with $\overline{X} < 18$ and $\overline{X} \gg 100$ because of the limited spatial allocations available.

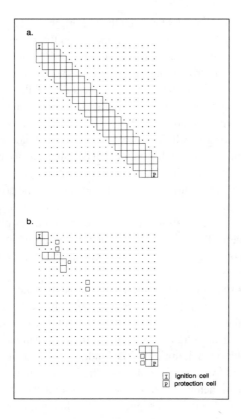

FIGURE 12.2

Binding burn path (a) and treatment areas (treatment amounts indicated by size of square for each cell) in optimal solution (b), with $\overline{X} = 18$ and one protection cell (20, 20).

Note that if the partial cell treatments are logistically problematic, this model can be solved with binary integer X variables that represent complete treatment (1) or no treatment (0) of each cell. We solved such a model that was otherwise identical to the one shown in figure 12.2 and obtained a solution with an objective function attainment within 3% of the bound (the solution in figure 12.2) in just a few minutes of computing time. The solution in figure 12.2 appears to be implementable, however, with two large treatment areas separated by more than 8 hours of burn time.

How would the solution in figure 12.2 compare to simpler spatial allocations? Are the model results sensitive to spatial layout or just to the amount of treatment allowed? A spatially neutral layout might be to simply place the 18

treated cells on the diagonal between the fire origin cell (1, 1) and the protection cell (20, 20). This solution was imposed on the model, and the ignition
time of cell (20, 20) decreased from that in figure 12.2 by more than 3 hours,
to 833 minutes. The critical burn path with this solution imposed is the diagonal of cells just above or below the treatment diagonal.

A treatment layout that might be applied in the absence of an optimization
analysis is shown in figure 12.3. Here, the 18 treated cells are all imposed
around the fire origin cell. With this strategy, the binding burn path cuts
straight through the area on the diagonal and ignites the protection cell at 875
minutes (2½ hours before the optimized solution in figure 12.2). The mirror
image of this strategy, clustering the treatment around the protection cell, was
also imposed with very similar results. This similarity highlights the assumption in our model formulation that fire acceleration takes place within cells,
not between them. The effect on protection cell ignition time is the same for
these two mirror-image solutions, even though one attacks the fire early and
the other late.

Figure 12.4 shows one other imposed treatment layout. Here, an area (still
18 cells) in the center of the grid is constrained to be treated in the model. The
result is to ignite the protection cell at 958 minutes, which is almost an hour
and a half later than in figure 12.3 but is still more than an hour sooner than
the optimal solution in figure 12.2. The binding burn path in figure 12.4 shows
that the imposed treatment layout slows the ignition time in cell (20, 20) by
bending and lengthening the binding path around the treatment area. The
results in figures 12.3 and 12.4 suggest that for this simple example, the mixed
strategy in the optimization solution has a definite advantage over strategies
that focus on a single area. If the landscape was heterogeneous, with more
variation in any of the model parameters, then the optimal solution could be
more complex and more difficult to anticipate. The solutions shown here
demonstrate a simple solution pattern that is not complicated by such landscape variation. Even with such a homogeneous example, however, the results
appear to be spatially sensitive.

Figure 12.5 shows an optimal solution with three protection cells, located
at (10, 20), (20, 10), and (20, 20). The objective function for this solution used
a MAXMIN (Hof et al., 1986) operator as follows:

Maximize

λ,

$\lambda \leq T^{\circ}_{10,20}$,   (12.10)

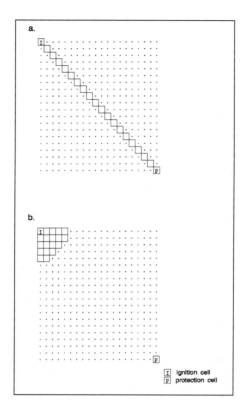

FIGURE 12.3

Binding burn path (a) and treatment areas (treatment amounts indicated by size of square for each cell) in imposed solution around the origin (b), with one protection cell (20, 20).

$$\lambda \leq T^{\circ}_{20,10}, \tag{12.11}$$

$$\lambda \leq T^{\circ}_{20,20}. \tag{12.12}$$

This approach maximizes the minimum ignition time across the three protection cells, an equity-oriented approach. The area treated ($\overline{X}$) was set at 100 cells for this solution. This area is optimally laid out in a manner similar to the hourglass strategy seen in figure 12.1 but split among the three protection areas. The binding burn path also splits among the protection areas, as seen in the top part of figure 12.5. This binding path confirms that the ignitions of all three protection cells are constraining in solution. The MAXMIN operator

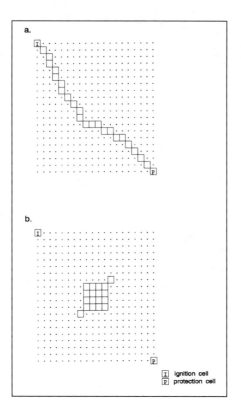

FIGURE 12.4

Binding burn path (a) and treatment areas (treatment amounts indicated by size of square for each cell) in imposed solution near the center (b) with one protection cell (20, 20).

causes treatment to be laid out to create this equitable arrangement. The ignition time for all three protection areas is 1228 minutes. This is 208 minutes sooner than in figure 12.1, reflecting the fact that the same treatment area (resource) is now being used to delay the ignition of three spatially separated protection areas instead of one. We also solved the model leaving inequality (12.12) out, including only two protection cells. Even with only two protection areas, the time of ignition (in both) was 1251 minutes, more than 3 hours sooner than in figure 12.1. It is also noteworthy that the optimal treatment strategy in figure 12.5 focuses more on the area around the origin and less around the protection cells than in figure 12.1. This reflects the fact that effort expended around the fire origin helps delay ignition of all three protection

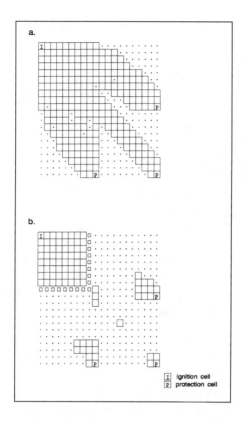

FIGURE 12.5

Binding burn path (a) and treatment areas (treatment amounts indicated by size of square for each cell) in optimal solution (b) with $\overline{X} = 100$, three protection cells (20, 10), (10, 20), and (20, 20), and a MAXMIN objective function.

areas, whereas effort expended around any of those areas delays only that one area's ignition. Even with this division of effectiveness, the optimal solution still allocates some treatment around each of the protection areas.

An alternative objective function with multiple protection areas would simply be a weighted sum of those areas' ignition times. This approach is not as equitable as the MAXMIN, even if equal weights are used (e.g., because the protection areas may be located with different distances from the fire origin). Figure 12.6 shows a solution, again with $\overline{X} = 100$ and the same three protection cells but maximizing the simple sum of the three ignition times. Cell (20, 20) ignites at time 1275, 67 minutes later than cells (10, 20) and (20, 10).

The average ignition time across the three cells (1230.33 minutes) is only 2.33 minutes later than the simultaneous ignition times in figure 12.5, but cells (10, 20) and (20, 10) ignite 20 minutes sooner than in figure 12.5. It is interesting that more attention is given to cell (20, 20) at the expense of the area around the fire origin rather than the areas around the other two protection areas. If different protection areas have different priorities, a weighted sum or a pre-emptive weighting approach might be appropriate. If it is not possible to differentiate the importance of different protection areas, then the MAXMIN approach probably is desirable. Other objective functions may also be useful, such as piecewise approximations of nonlinear forms, but the simple ones demonstrated here are the easiest to apply.

In one last experiment with the model (reported in figure 12.6), we imposed a solution with .25 of all cells treated (all X_{ij} = .25). The three protection cells were all ignited at 979 minutes, which is 251 minutes sooner than the average ignition time in figure 12.6. This is simply another demonstration that the solution is spatially sensitive. The binding burn path in this solution (not shown) was one cell wide and followed the diagonal to the center of the grid, then split in the directions of the three protection cells.

Extensions

The model formulation clearly includes some important assumptions. Most importantly, we assume that timing of fire spread and ignition of specific protection areas define the objective of the fire management effort being allocated by the model. Fire behavior is treated (deterministically) as being predictable, and management treatment is assumed to slow the spread of the fire front. Perhaps our most critical assumption with regard to fire behavior is that the movement of the fire front is predictable cell by cell, with no cell-to-cell interaction (such as larger-scale buildup of velocity from cell to cell). The most obvious extensions to this model would be formulations that allow relaxation of these critical assumptions. A more fundamental extension of this work would involve a different problem.

The orientation of the chapter up to this point has been the immediate management of a fire after it has started, but under certain circumstances our modeling approach might be applicable to long-term fuel management. If the objective of long-term fuel management is to mitigate the effects of a particular target fire with known origin and spread behavior, then our approach could be adapted as follows.

First, the management variables must be defined as applications of fuel reduction efforts in each cell (e.g., prescribed burning, mechanical removal),

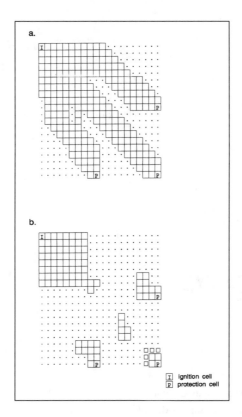

FIGURE 12.6

Binding burn path (a) and treatment areas (treatment amounts indicated by size of square for each cell) in optimal solution (b) with $\overline{X} = 100$, three protection cells (20, 10), (10, 20), and (20, 20), and an equally weighted summation objective function.

and these efforts must be scheduled over a fairly long period of time because only so much fuel reduction can be accomplished in a given year or season. Thus discrete time periods of, say, 1–10 years would be defined and indexed with t. The trajectory of each cell's fuel load over time and its response to fuel management would have to be tracked as well. Such a model might be formulated as follows:

Maximize

$$\lambda \qquad\qquad (12.13)$$

subject to

$$\lambda \le T^{\circ}_{mnt} \qquad \forall t,$$

$$T^{\circ}_{abt} = 0 \qquad \forall t, \qquad (12.14)$$

$$T^{\circ}_{ijt} \le T'_{hkt} \qquad \forall t \qquad (12.15)$$
$$\forall (h, k) \in \Omega_{ij}$$
$$\forall i, j,$$

$$T'_{ijt} - T^{\circ}_{ijt} = f_{ij}(F_{ijt}) \qquad \forall i, j, t, \qquad (12.16)$$

$$F_{ijt} = \sum_{k=1}^{K_{ij}} B_{ijkt} X_{ijk} \qquad \forall i, j, t, \qquad (12.17)$$

$$\sum_{k=1}^{K_{ij}} X_{ijk} \le 1 \qquad \forall i, j, \qquad (12.18)$$

$$\sum_{i} \sum_{j} \sum_{k=1}^{K_{ij}} D_{ijkt} X_{ijk} \le \overline{X}_t \quad \forall t, \qquad (12.19)$$

$$0 \le X_{ijk} \le 1 \qquad \forall i, j, k \qquad (12.20)$$

Indexes

i indexes cell rows, as does h,
j indexes cell columns, as does k,
k indexes management prescriptions (there are K_{ij} of them for cell ij).

Variables

T°_{ijt} = the time at which the target fire front ignites cell ij if it occurs in time period t,

T'_{ijt} = the time at which the target fire front leaves cell ij if it occurs in time period t,

T'_{hkt} = the time at which the target fire front leaves cell hk (and potentially ignites cell ij) if it occurs in time period t,

F_{ijt} = the fuel available for combustion in cell ij and in time period t,

X_{ijk} = the proportion of cell ij allocated to management prescription k,

f_{ij} = an empirical function, as defined previously.

Parameters

a, b, m, n, Ω_{ij} as defined previously,

B_{ijkt} = the available fuel in time period t, in cell ij, that results if management prescription k is applied,

D_{ijkt} = a dummy parameter that is equal to 1 if management prescription k applies fuel reduction in time period t, in cell ij, and is zero otherwise,

$\bar{X}_t$ = the total number of cells that can be treated with fuel reduction in time period t.

The objective function (12.13) maximizes the minimum ignition time of the protection cell mn across time periods (the target fire might occur in any time period). Other objective functions that aggregate time periods would also be possible. Constraints (12.14)–(12.16) function like (12.2)–(12.4) except that a potential target fire in each time period is accounted for. Constraint (12.17) applies the management variables that now include scheduling to determine the changing available fuel load in each cell ij over time. Constraint (12.18) restricts the sum of the management prescription variables to be less than or equal to 1 for each cell, and constraint (12.19) limits the amount of fuel management treatment that can be applied in each time period (analogous to (12.7) in the previous model). As with the previous model, it may be desirable to replace (12.20) with binary integer constraints on all X_{ijk} to force either complete treatment or no treatment of each cell in any given time period. As the formulation stands, it is assumed that the B_{ijkt} parameter is applicable to fractional values of X_{ijk}.

In our model for applying management effort during a fire, the linkages between modeled delay efforts, logistic delivery systems, and containment efforts could be quite important. In our model for applying management effort through fuel reductions over a period of years or decades, the application logistics are far less demanding and fire containment is irrelevant, but the reliance on a single target fire for spatially and temporally planning treatments is an important limitation. The use of a target fire as a planning objective is an approach similar to the use of particular stormflow events in chapter 3. However, highly random fire behavior conditions such as wind speed and direction or points of origin are far less predictable than the path of water flow on a landscape. If a different fire event ultimately occurs, the fuel management strategy based on the target fire may or may not be desirable. New approaches to defining the objective seem to be needed for prefire fuel treatment planning.

The case example presented here does not reflect any particular management setting, and no generalizations are intended. The purpose of our stylized example was to demonstrate the model formulations with easily interpretable results. As we have discussed, application of models such as ours relies on substantial capability in predicting fire behavior. Constructing and solving optimization models is much less of a problem because the entire process takes only a few minutes once the data set is in place. Of more concern is that our model depends on predictions of processes that are not completely understood and are subject to random, perhaps chaotic or catastrophic behavior. This chapter is just an exploratory look at the use of timing-variable optimization models for fire and fuel management. In part IV we use optimization models for more theoretical ecological investigations.

PART IV

USING OPTIMIZATION TO DEVELOP
HYPOTHESES ABOUT ECOSYSTEMS

Up to this point, we have treated optimization models as prescriptive devices: tools to help choose the most efficient way to allocate certain scarce resources. Our optimization models obviously contain a large component that could be regarded as simulation to predict the ecosystem response to alternative management actions. However, it is possible that ecosystems actually optimize themselves. In this case, optimization could be used to simulate ecosystem behavior per se. The reaction–diffusion simulations of chapter 6, for example, produce identical results over time and converge to the same equilibria as do optimization solutions from models using the same (fixed) landscapes and parameter sets. When modeled ecosystems are attracted to equilibria of one sort or another, optimization models could be used to calculate (simulate) that equilibrium.

At the smaller biological scale of the individual, investigators have suggested the possibility that plants exhibit optimizing behavior (Bazzaz and Grace 1997). At the population or metapopulation scale, biodiffusion models such as Allen's (1987), which we based part II on, are themselves based on an assumption of limited maximizing behavior by animal populations. At the community or ecosystem scale, the process of natural selection is certainly an optimization process in the sense that it eliminates inefficient components in the system. It also seems reasonable that the ecosystem components that remain do so in part because they function efficiently. Given enough time, the system that survives might then be expected to exhibit optimizing behavior as well.

In part IV we provide two examples of how optimization models can be used to simulate the behavior of ecosystems and thereby be used to develop testable hypotheses about ecosystem behavior. In chapter 13, we use a large linear program much like those developed in part II to develop hypotheses with regard to limiting factors operating at multiple spatial scales. The model in this chapter uses optimization to simulate an equilibrium, created by "attractors." Chapter 14 focuses on the organism scale and develops two alternative models for the allocation of carbon in trees. Thus this chapter treats trees as efficient allocators of resources, with optimizing behavior. Calculus is used to develop testable hypotheses for confirming (or not confirming) each model. A numerical model is then analyzed, which suggests that if one model is supportable, the other probably is not. Chapter 14 is the only chapter in the book in which we use nonlinear programming. We still regard it as a pragmatic example because the problem as defined calls for only a small model, which appears to be convex. When the purpose of optimization is to simulate ecological processes and develop theoretical hypotheses, the standards for being pragmatic may be different from those in the prescriptive context. Chapter 14 is also the only chapter in the book that does not involve the landscape scale.

13

MULTISCALED ECOLOGICAL LIMITING FACTORS

Although we have modeled carrying capacity for single populations as a simple equilibrium or saturation point, the concept is widely recognized as being more complex (Dhondt 1988). For more than a century, biogeographers have recognized that population distributions tend to be limited by a multitude of factors (Haeckel 1866; Shelford 1911; Udvardy 1969). These factors also affect population abundance (as in chapter 10) and can affect populations at different spatial scales (Morris 1987; Levin 1992). For example, many species forage over spatially extensive areas but reproduce in specific locales because of obligate relationships with habitat characteristics that occur sporadically across a landscape (Orians and Wittenberger 1991). Under such circumstances, populations can be limited at microscales by the availability of suitable breeding sites and at macroscales by food availability. Similar instances of multiscale population limitation are found in predator–prey systems, where species can be limited broadly by wide-ranging predators or locally by the availability of breeding sites (Henschel and Lubin 1997). Less familiar examples can be found among species characterized by multiphase life cycles. For example, many adult forms of marine intertidal organisms are sessile and subject to population limitations stemming from the local availability of suitable substrate. Conversely, the larval phase of these species often is highly mobile, distributed diffusely throughout the marine environment, and subject to large-scale limitations caused by nutrient availability, predation, or weather (Roughgarden et al. 1988).

This chapter was adapted from M. Bevers and C. H. Flather, The distribution and abundance of populations limited at multiple spatial scales, *Journal of Animal Ecology* 68 (1999): 976–987, with permission from the publisher, the British Ecological Society.

Indeed, there is a general recognition that observed population distribution and abundance patterns result from limitations at several spatial scales. As Levin (1992:1943) puts it, "in some cases, the patterns must be understood as emerging from the collective behaviors of large ensembles of smaller scale units. In other cases, the pattern is imposed by larger scale constraints." This chapter focuses on the combined effect of such multi-scale limitations.

Ecological adaptations to multiscale influences on populations are numerous (Hengeveld 1987; Menge and Olson 1990), resulting in a variety of distinct simulation approaches for particular populations. Given this variation, general theoretical treatment of how multiscale limiting factors can affect population distribution and abundance should be useful. This chapter reports the results of an investigation of combined micro- and macroscale limiting factor effects from Bevers and Flather (1999a). As in part II, we (Bevers and Flather) used linear programming with a reaction–diffusion formulation (again see Okubo 1980; Holmes et al. 1994) to explore these effects for a theoretical population randomly dispersing throughout a single patch of habitat.

In chapter 6, population arrangements converged to contiguous Gaussian-like spatial distributions of abundance near extinction thresholds, whereas in chapter 7 a fragmented population was observed as an unanticipated optimal solution. These observations led us to hypothesize that multiscale limiting factors will allow fragmented populations to occur under a broad set of reaction–diffusion parameters, converging toward Gaussian-like population clusters as critical thresholds are approached. We further examine population clustering in heterogenous habitat by imposing local habitat preferences within a patch.

Formulation

Let us start again (as in the introduction to part II) with the model from Allen (1987, following Levin 1974) converted from patches to describe a random walk reaction–diffusion model for a complex of N habitat cells as

$$\frac{dv_i}{dt} = v_i f_i(v_i) + \sum_{j=0}^{N} D_{ij}(v_j - v_i) \quad i = 1, \ldots, N, \tag{13.1}$$

$$D_{ij} = D_{ji} \geq 0 \qquad \qquad \forall i, j,$$

$$v_0 = 0,$$

where v_i is the adult population in cell i as a function of time t, and $f_i(v_i)$ is the per capita rate of reproduction (net of mortality unassociated with dispersal, as discussed later). D_{ij} is a cell-to-cell passive diffusivity constant that determines the net cell i population gain from (or loss to) cell j based on the difference between the cell populations. Nonhabitat is indexed by $j = 0$. In this model, dispersing organisms can successfully traverse regions of nonhabitat, but some of them perish and are treated as dispersers into nonhabitat. The nonhabitat population v_0 is fixed at zero so that D_{i0} times $-v_i$ defines a population loss caused by unsuccessful dispersal from any cell i. In this chapter, we use a 13×13 block of square cells to examine population distribution and abundance across 169 breeding sites in a single habitat patch.

Reproduction and dispersal are modeled as simultaneous processes in equation (13.1) so that reproduction in each cell can implicitly contribute to the diffusion summation term. In this unstructured model, only adults and offspring are distinguishable. For simplicity, we assume that reproduction occurs at a constant per capita rate (i.e., $f_i(v_i) = r$ for all i) and that adults and juveniles disperse identically.

To estimate cell-to-cell dispersal, we again define a parameter g_{ij} as the proportion of organisms from cell i expected to disperse to cell j (by any of many routes) per unit of time. With adults and juveniles dispersing identically, all diffusivity constants D_{ij} are equal to $(1 + r)g_{ij}$. Again for simplicity, we assume that diffusion from each cell in the patch is identical and that organisms disperse, on average, from the center of each cell outward in uniformly random directions. Dispersal probabilities decline with distance (x) according to

$$p_X(x) = \mu^{-1} \exp\left[-(x - \theta)/\mu\right] \quad x > \theta, \mu > 0,$$

as defined by a mean dispersal distance (μ) from the center of the home cell, using a minimum dispersal distance (θ) of zero. Diffusion proportions (g_{ij}) are estimated by numerical approximation over distances and angles defined by the boundaries of each destination cell (indexed by j) relative to the center of each source cell (indexed by i). Using an exponential distribution for dispersal distance (as in Fahrig 1992) results in a globally connected cellular lattice.

Case Example

We want to represent a microscale limit such as breeding site capacity and a macroscale limit such as a predation effect in a general way. Our two spatial scales are the cells and the patch. To apply a microscale capacity limit, we add to the model a set of constraints limiting the adult population in each cell:

$$v_i \le b_i \quad \forall i. \tag{13.2}$$

For most of our analyses a homogeneous patch is assumed, and the cellular breeding site capacity parameter b_i is set to a scale value of 1.0 population unit for all cells. To apply a macroscale capacity, another constraint limiting total patch population is added to the model:

$$\sum_i v_i \leq C, \tag{13.3}$$

where C is the landscape-level or total patch capacity.

The generality and simplicity of equations (13.1)–(13.3) eliminate many ecological details from the model. For example, our use of fixed cell sizes combined with equation (13.2) disregards fluctuation in territory sizes, the possibility of overpopulation, or cell occupancy by nonbreeding adults. Instead, any surplus cellular population is presumed to perish, and all adults are assumed to reproduce. Our use of a constant r value for reproduction in equation (13.1) is consistent with that but is simplistic as well. Combined with our choice of diffusion parameters, however, these assumptions produce a probability of dispersers surviving that is proportional to the amount of unoccupied habitat surrounding a given cell, with distance decay. Similarly, equation (13.3) ensures that any surplus patch population also perishes because of a macroscale limiting factor. With this approach we can focus on identifying the range of suitable population arrangements without having to specify the behavioral or energetic mechanisms underlying the limiting factors. The effects of such mechanisms are considered in this chapter's Discussion.

Model population distribution and abundance under equilibrium conditions can be observed by setting the rates of change (the dv_i/dt terms in equation 13.1) to zeroes and maximizing total population with equations (13.1)–(13.3) as constraints. We recognize that random perturbations can disrupt any balance in the system, causing a continuing sequence of disequilibria. Nonetheless, we are interested in equilibrium states because in the aggregate those that maximize total population form the attractors (Çambel 1993) toward which the system tends to move after being perturbed (Cushing et al. 1998). Optimization (linear programming in this case) provides a convenient method for finding those equilibria.

Because a variety of maximal equilibria may exist for our 13×13 cell patch under any given set of parameter values, a systematic approach is needed to explore the possible population arrangements. Our approach involves solving the model (equations 13.1–13.3) with four different objective functions: CNTR, EDGE, FRAG, and EVEN (defined later). Each of these objective functions allows us to search systematically for extreme distribution and abundance patterns while maintaining maximum equilibrium population conditions for a given set of parameter values. Following a preliminary optimization using

Maximize

$$\sum_{i=1}^{N} v_i \qquad (13.4)$$

with equations (13.1)–(13.3) as constraints to identify the maximum equilibrium population, we convert the landscape-level capacity constraint (equation 13.3) to an equality constraint and set that right-hand side (C) to the new equilibrium population level just determined, forcing the model to maintain it. We then reoptimize the problem four more times, replacing equation (13.4) with a different distribution-favoring objective function each time.

In our 13×13 patch, the CNTR objective encourages high population densities in the center of the patch by maximizing a weighted sum of the cellular populations as follows:

Maximize

$$\sum_{i=1}^{N} w_i v_i, \qquad (13.5)$$

with the greatest weight placed on the center cell ($w_{85} = 1.0$, with cells indexed from the upper left corner). Cell weights are reduced by a factor of 10 in each successive layer moving outward from the center cell, so that the outermost cells are weighted 1.0×10^{-6}. The EDGE maximization objective function is weighted in the opposite manner to encourage high population densities along the edge of the complex, with the center cell weighted at 1.0 and the outermost cells each weighted at 1.0×10^6. The FRAG maximization objective function encourages hyperdispersion of cellular populations, with weights of 1.0 for each of the 36 cells occurring in even-numbered rows and columns in the complex (so that none of these 36 cells adjoin). Cells with odd-numbered rows or columns, but not both, directly facing those 36 cells are assigned weights of -1.0×10^6. Cells with both odd-numbered rows and columns diagonal to the 36 positively weighted cells are assigned weights of -1.0×10^3. The EVEN objective function uses a MINIMAX (Luce and Raiffa 1957) formulation (equation 13.6), similar to the MAXMIN operator used in chapter 10, to encourage evenly distributed equilibrium cellular populations:

Minimize

$$\lambda \tag{13.6}$$

subject to

$$\lambda \geq v_i \quad \forall i,$$

where the largest equilibrium cellular population (λ) is minimized.

When only a single distribution and abundance pattern maximizes the total population, all four of these objective functions are forced to produce the same equilibrium population arrangement. Conversely, solutions from our four objective functions are different when there is flexibility in how populations can be arranged on the landscape (i.e., multiple equilibria exist that can support the same total population). For each solution from these objective functions, we calculate a pair of contiguity measures (Geary 1954; Dacey 1968) to describe the degree to which population distribution is clustered (referred to as z_{11}) and abundance is concentrated (referred to as K) on the landscape (see the Appendix for definition and interpretation of these measures).

Results

We begin by examining equilibrium population distributions for a homogenous habitat patch 13×13 units in size so that each cell measures one distance unit per side. Cellular population capacities (b_i) are set to 1.0 (perhaps representing many organisms) for all 169 cells. The net reproduction rate (r) is set to .85, and mean dispersal distance (μ) is set to 2.5 units of distance. This set of values represents our base parameter settings. Figure 13.1 shows two of several alternative equilibrium population distributions found by random trials with the patchwide capacity (C) set to 36 population units. Although both maps exhibit substantial clustering ($z_{11} > 3.31$), it is clear from figure 13.1a that small, isolated cellular populations can occur under equilibrium conditions. Figure 13.1a also shows a more concentrated pattern of abundance ($K = .57$), whereas figure 13.1b displays a more even abundance ($K = .27$), resulting from substantially fewer unoccupied cells.

Despite the similarities and differences between figures 13.1a and 13.1b, inferences are difficult without more systematic analysis. At these parameter settings, the objective function (equation 13.4) is limited to a total equilibrium population of 36 by the complexwide capacity constraint (equation 13.3). To examine the flexibility possible in equilibrium population arrangements, we converted equation (13.3) to an equality constraint with C set to 36 and

a.

0.45	0.65	0.75	0.72	0.56								
0.54	0.83	0.99	0.97	0.75						0.04		
	0.78	1.00	1.00	0.80								
			1.00				0.20			0.07		
	0.73	0.99	1.00	0.91	0.63					0.10		
	0.70	0.99	1.00	0.87	0.61			0.18				
		0.81	0.85					0.23	0.24			
	0.42	0.59	0.61	0.47					0.39	0.37		
					0.21			0.02	0.53	0.51		
0.15								0.53	0.67	0.64	0.45	
		0.12			0.16		0.38	0.59	0.71	0.67	0.49	
								0.52	0.63	0.59	0.42	
								0.37	0.44	0.40		

$Z_{11} = 8.89$

$K = 0.57$

b.

0.21	0.27	0.28	0.28	0.27	0.22		0.25	0.28	0.28	0.28	0.26	0.20
0.28	0.28	0.28	0.28	0.28			0.28	0.28	0.28	0.28	0.28	0.26
0.28	0.28	0.28	0.28	0.28			0.28	0.28	0.28	0.28		0.28
0.28	0.28	0.28	0.28	0.28				0.28	0.28	0.28	0.28	0.28
0.28	0.28	0.28	0.28	0.28				0.12	0.28	0.28	0.28	0.28
0.28	0.28	0.28			0.28	0.28			0.28	0.28	0.28	0.28
0.28	0.28	0.28	0.28	0.28	0.28			0.28	0.28	0.28	0.28	0.28
0.28	0.28				0.28	0.28	0.28		0.28	0.28	0.28	0.28
0.28	0.28	0.28	0.28	0.28	0.28	0.28	0.28	0.28	0.28	0.28	0.28	0.28
0.28	0.28	0.28	0.28					0.28	0.28	0.28	0.28	0.28
0.23	0.28	0.28	0.28			0.28		0.28	0.28	0.28	0.28	0.26
	0.20			0.27	0.28	0.28	0.28	0.28	0.28	0.28	0.28	0.28
		0.22	0.27	0.28	0.28	0.28	0.28	0.28	0.28	0.28	0.28	0.21

$Z_{11} = 3.32$

$K = 0.27$

FIGURE 13.1

Two of many possible equilibrium population arrangements in a 13 × 13 cell habitat patch with parameter settings $C = 36$, $r = .85$, $\mu = 2.5$, and all $b_i = 1.0$.

replaced equation (13.4) with the CNTR, EDGE, FRAG, and EVEN objective functions (one at a time). The resulting population arrangements (still at equilibrium levels of 36) are shown for these objective functions in figures 13.2a–13.2d, respectively. Each of these figures portrays one solution from a small family of possible arrangements that vary little (e.g., mirror images are always alternative solutions for our symmetric landscape).

The reaction–diffusion constraint (equation 13.1) appears to have little effect on the CNTR objective function results. The central 25 cells are fully populated and the remaining 11 population units are arranged more or less haphazardly around the next concentric layer of cells. However, equation (13.1) does appear to have some effect with the EDGE objective function. Instead of placing all 36 population units in the outermost set of cells, the equilibrium population is distributed throughout the two outer concentric rings and in the corners of the next ring inward. The model apparently is unable to populate the edge of the complex without some connecting cellular populations toward the interior. Similarly, the FRAG objective function results show that the 36 positively weighted (hyperdispersed) cells cannot be used to contain all 36 population units given passive reaction–diffusion processes at these parameter settings. All the diagonally adjoining cells are populated despite negative weights in the objective function. Six of the adjacent cells near the center of the complex are populated despite very negative weights in the objective function. The EVEN objective function results, on the other hand, appear to be largely unconstrained by reaction–diffusion processes, similar to

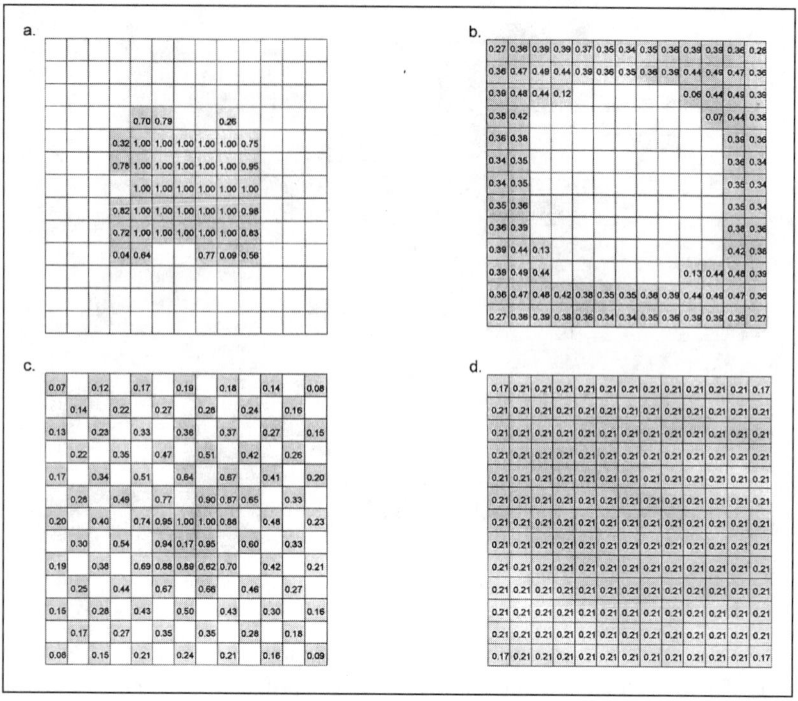

FIGURE 13.2

Equilibrium population arrangements in a 13×13 cell habitat patch with parameter settings $C = 36$, $r = .85$, $\mu = 2.5$, and all $b_i = 1.0$ for the (a) CNTR, (b) EDGE, (c) FRAG, and (d) EVEN objective functions.

the CNTR objective function results. Only the four corner cells of the complex are affected in this case. Contiguity statistics for these results are reported in the first line of table 13.1. Overall, the model appears to have a large degree of flexibility in equilibrium population arrangement at these parameter settings, although that flexibility is somewhat limited. Because of reaction–diffusion constraints, some amount of clustering seems unavoidable in all of these solutions, but the clusters are not necessarily prominent (e.g., figure 13.2c).

Reproduction and Dispersal Effects

As discussed in chapter 6, Skellam (1951) and others have demonstrated that for continuous space and time models of theoretical populations with passive

TABLE 13.1

Contiguity Statistics for Population Distribution (z_{11}) and Concentration (K) Under the CNTR, EDGE, FRAG, and EVEN Objective Functions for a 13 × 13 Cell Habitat Patch with Base Parameter Settings $C = 36$, $r = .85$, $\mu = 2.5$ and All $b_i = 1.0$, and for Individual Changes to Those Parameter Settings

Parametric Change	CNTR		EDGE		FRAG		EVEN	
	z_{11}	K	z_{11}	K	z_{11}	K	z_{11}	K
None (Base)	18.45	.857	8.54	.371	−.01	.396	—[a]	.213
$r = .5276$	19.34	.750	7.42	.279	1.87	.364	—[a]	.213
$r = .20515$	—[a]	.213	—[a]	.213	—[a]	.213	—[a]	.213
$\mu = 4.7661$	18.52	.468	5.66	.250	3.55	.327	—[a]	.213
$\mu = 7.0322$	—[a]	.213	—[a]	.213	—[a]	.213	—[a]	.213
$b_{85} = 2.0$	19.72	.942	8.56	.375	.12	.391	—[b]	—[b]
$v_{85} \geq 1.0$	19.72	.942	4.49	.359	.12	.391	—[b]	—[b]
$v_{85} \geq 1.5721$	19.72	.942	18.95	.918	19.60	.918	—[b]	—[b]

[a]The distribution statistic (z_{11}) cannot be computed when all cells in the complex are occupied.
[b]The MINIMAX form of the EVEN objective function can be ineffective when a lower bound is placed on some v_i, so no results are reported.

diffusion processes in a single patch of homogenous habitat, a critical patch size generally exists, below which extinction is expected. Critical extinction thresholds also exist in discrete models such as ours, as we showed in chapter 6 (see also Lande 1987 and Pagel and Payne 1996 for related demographic equilibrium extinction thresholds). In our model, this threshold is determined by a combination of habitat size and shape, the intrinsic rate of population growth (r), and mean dispersal distance (μ, hence all g_{ij}). By varying one of these parameters and holding the others constant, we can determine a critical set of threshold values for all parameters. Below the threshold point, the optimization model in equations (13.1)–(13.4) cannot produce any expected equilibrium population for the patch (i.e., equation 13.4 is maximized at zero, implying extinction). For a given habitat patch, decreasing r or increasing μ tends to push the equilibrium population toward (or below) extinction threshold conditions. Thus we can investigate more restrictive reaction–diffusion systems by decreasing r or increasing μ.

Returning to our base parameter settings with $C = 36$, all $b_i = 1.0$, $\mu = 2.5$, and $r = .85$, we decreased the net reproduction rate r value until we encountered an extinction threshold at $r = .20515$. With the r value set to .20514 or less, equation (13.4) is maximized at zero. Conversely, with the r value set to .20515, equation (13.4) is maximized at a total equilibrium population of 36 units because the complexwide capacity constraint (equation 13.3) is simultaneously limiting. With equation (13.3) converted to an equality constraint at $C = 36$, the CNTR, EDGE, FRAG, and EVEN objective functions (as well as equation 13.4) all produce identical Gaussian-like equilibrium population distributions (figure 13.3). No latitude exists in how the equilibrium population can be arranged when it is near extinction conditions, and our contiguity indices are invariant (table 13.1).

At $r = .5276$, about midway between our original r value and the extinction threshold r value, equilibrium arrangements from the CNTR, EDGE, FRAG, and EVEN objective functions (figures 13.4a–13.4d) appear to be moderate in flexibility (compare with figures 13.2a–13.2d and 13.3). As shown in table 13.1, with this r value we observe population clustering for the CNTR, EDGE, and EVEN objective functions, comparable to the results from our base parameter settings but with slightly lower concentrations. Population contiguity under the FRAG objective function deviates from the base results because the more restrictive reaction–diffusion system forces more of the population toward the center of the complex (see figure 13.4c).

Returning again to our base parameter settings with $C = 36$, all $b_i = 1.0$, $r = .85$, and $\mu = 2.5$, we next increased the mean dispersal distance until we encountered another extinction threshold at $\mu = 7.0322$. Again, the resulting equilibrium population is 36 because the complexwide capacity constraint (equation 13.3) and the reaction–diffusion constraints (equation 13.1) are simultaneously limiting. As with a critical r value, all the objective functions produce identical equilibrium population distributions (figure 13.5), and no latitude in arrangement exists. Considered as a three-dimensional graph, figure 13.5 is only slightly flatter than figure 13.3 despite much more equitable dispersal probabilities between cells. Setting μ midway between our initial value and the extinction threshold value produced a set of responses similar to figure 13.4, for which an intermediate r value was used. Changes in contiguity statistics are similar to those observed when we let r approach an extinction threshold, except that we observe greater population clustering for intermediate values of μ under the FRAG objective function (table 13.1). This suggests that the population becomes somewhat less tolerant of fragmented arrangements as μ approaches an extinction threshold, relative to r.

0.06	0.09	0.11	0.13	0.14	0.15	0.15	0.15	0.14	0.13	0.11	0.09	0.06
0.09	0.12	0.15	0.18	0.20	0.21	0.21	0.21	0.20	0.18	0.15	0.12	0.09
0.11	0.15	0.19	0.22	0.25	0.26	0.26	0.26	0.25	0.22	0.19	0.15	0.11
0.13	0.18	0.22	0.26	0.29	0.30	0.31	0.30	0.29	0.26	0.22	0.18	0.13
0.14	0.20	0.25	0.29	0.31	0.33	0.34	0.33	0.31	0.29	0.25	0.20	0.14
0.15	0.21	0.26	0.30	0.33	0.35	0.36	0.35	0.33	0.30	0.26	0.21	0.15
0.15	0.21	0.26	0.31	0.34	0.36	0.36	0.36	0.34	0.31	0.26	0.21	0.15
0.15	0.21	0.26	0.30	0.33	0.35	0.36	0.35	0.33	0.30	0.26	0.21	0.15
0.14	0.20	0.25	0.29	0.31	0.33	0.34	0.33	0.31	0.29	0.25	0.20	0.14
0.13	0.18	0.22	0.26	0.29	0.30	0.31	0.30	0.29	0.26	0.22	0.18	0.13
0.11	0.15	0.19	0.22	0.25	0.26	0.26	0.26	0.25	0.22	0.19	0.15	0.11
0.09	0.12	0.15	0.18	0.20	0.21	0.21	0.21	0.20	0.18	0.15	0.12	0.09
0.06	0.09	0.11	0.13	0.14	0.15	0.15	0.15	0.14	0.13	0.11	0.09	0.06

FIGURE 13.3

Equilibrium population arrangements in a 13 × 13 cell habitat patch near extinction threshold conditions with parameter settings $C = 36$, $r = .20515$, $\mu = 2.5$, and all $b_i = 1.0$ for all objective functions.

Heterogeneous Habitat Effects

In our investigations so far, we have considered only homogeneous habitat patches (all $b_i = 1.0$). Our observations—that flexibility in equilibrium population distribution and abundance decreases as reaction–diffusion processes approach extinction threshold conditions—could be an artifact of using a simple homogeneous environment. To explore possible contiguity effects in a heterogeneous environment, we increased the saturation capacity of just the center cell in our 13 × 13 cell patch from one population unit to two ($b_{85} = 2.0$). One set of results from maximizing equation (13.4) with $C = 36$, all other $b_i = 1.0$, $r = .85$ and $\mu = 2.5$ is shown in figure 13.6a, although many equilibrium distributions are possible (contiguity statistics for the CNTR, EDGE, FRAG, and EVEN objective functions are reported in table 13.1). The distribution in figure 13.6a is noteworthy, however, because no equilibrium population occurs in the center cell. Without behavioral or energetic mechanisms in the model, passive reaction–diffusion (equation 13.1) does not necessarily cause the solution to exploit the higher-quality habitat represented by increases in breeding site capacity. Some form of preferential habitat selection has to be included in the model if we want equilibrium populations to occupy (prefer) certain habitat areas.

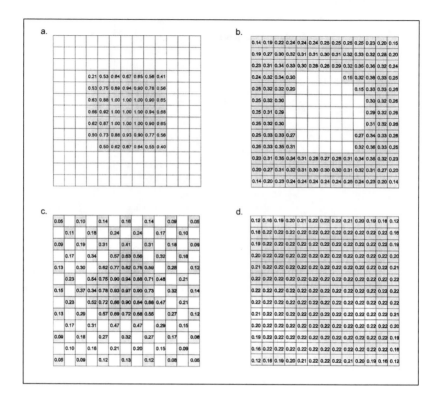

FIGURE 13.4

Equilibrium population arrangements in a 13 × 13 cell habitat patch with net reproduction $r = .5276$ (approximately midway between initial settings [figure 13.2] and extinction threshold conditions [figure 13.3]) and unchanged values for all other parameters ($C = 36$, $\mu = 2.5$, and all $b_i = 1.0$) for the (a) CNTR, (b) EDGE, (c) FRAG, and (d) EVEN objective functions.

Biasing dispersal probabilities to produce lower emigration rates from some sites would be a reasonable way to model preferential habitat selection but would also cause the results to be less comparable with those of our previous experiments. Without deviating from passive diffusion, a simple way to model cellular preferences is to place lower bounds on the cellular population (v_i) variables, requiring some population to occupy those cells. Small lower bounds close to zero imply weak habitat preferences. Large lower bounds close to saturation capacity (b_i) imply stronger habitat preferences. Figure 13.6b shows one arrangement resulting from setting a lower bound of 1 unit on the population in the center cell ($1 \leq v_{85} \leq b_{85}$, with b_{85} still set to 2). As

0.10	0.12	0.14	0.15	0.16	0.17	0.17	0.17	0.16	0.15	0.14	0.12	0.10
0.12	0.15	0.17	0.19	0.20	0.21	0.21	0.21	0.20	0.19	0.17	0.15	0.12
0.14	0.17	0.20	0.22	0.24	0.25	0.25	0.25	0.24	0.22	0.20	0.17	0.14
0.15	0.19	0.22	0.25	0.26	0.27	0.28	0.27	0.26	0.25	0.22	0.19	0.15
0.16	0.20	0.24	0.26	0.28	0.29	0.30	0.29	0.28	0.26	0.24	0.20	0.16
0.17	0.21	0.25	0.27	0.29	0.30	0.31	0.30	0.29	0.27	0.25	0.21	0.17
0.17	0.21	0.25	0.28	0.30	0.31	0.31	0.31	0.30	0.28	0.25	0.21	0.17
0.17	0.21	0.25	0.27	0.29	0.30	0.31	0.30	0.29	0.27	0.25	0.21	0.17
0.16	0.20	0.24	0.26	0.28	0.29	0.30	0.29	0.28	0.26	0.24	0.20	0.16
0.15	0.19	0.22	0.25	0.26	0.27	0.28	0.27	0.26	0.25	0.22	0.19	0.15
0.14	0.17	0.20	0.22	0.24	0.25	0.25	0.25	0.24	0.22	0.20	0.17	0.14
0.12	0.15	0.17	0.19	0.20	0.21	0.21	0.21	0.20	0.19	0.17	0.15	0.12
0.10	0.12	0.14	0.15	0.16	0.17	0.17	0.17	0.16	0.15	0.14	0.12	0.10

FIGURE 13.5

Equilibrium population arrangements in a 13×13 cell habitat patch near extinction threshold conditions with parameter settings $C = 36$, $r = .85$, $\mu = 7.0322$, and all $b_i = 1.0$ for all objective functions.

in figure 13.6a, some clustering is evident, but several cellular populations remain small and isolated. Although all equilibrium populations now include a population of at least 1 unit in the center cell, many alternative distributions are still possible, and most population contiguity measures are identical to the case with no preferential habitat selection (table 13.1). The EDGE objective function is the one exception. Two population clusters occur rather than one, causing a large drop in z_{11} because the lower bound restriction on v_{85} forces some population in and around the center cell while the objective function locates the remaining population around the edge of the complex.

As we increase the lower bound on the equilibrium population in the center cell, population arrangement alternatives decrease until at a lower bound of 1.5721 units (figure 13.6c) little flexibility is left. All objective functions show high contiguity statistics as the population clusters around the preferred habitat (table 13.1). Given the other parameter settings, it is impossible to increase v_{85} further even though the cellular capacity limit has not been reached (maximum $v_{85} = 1.5721 < b_{85} = 2$). The reaction–diffusion constraints (equation 13.1) are strongly limiting, forcing the entire equilibrium population to cluster around the center cell to support the largest population possible in the preferred cell under the circumstances.

To test for this effect at the edge of the habitat patch, we reset the lower bound on the population in the center cell back to zero and gradually increased

a.

0.06	0.12	0.18										
	0.22	0.33	0.47	0.60	0.65	0.57	0.38			0.08		
		0.23	0.77	0.89	0.76							
		0.57	0.26	1.00	0.88							
			0.88	1.00	0.95			0.14				
			0.88	1.00	0.99							
			0.79		1.00	0.81		0.17				
			0.77		1.00	0.90						
			0.78	0.99	0.66	0.87	0.58					
	0.29				1.00	1.00						
	0.45	0.69	0.94	1.00	0.79							
			0.32	0.86	0.74		0.34	0.21				
	0.29	0.42	0.55	0.63	0.57	0.43	0.30					

b.

0.15	0.12	0.28	0.31	0.26								
0.18		0.38	0.43	0.37								
0.30	0.41	0.46	0.41			0.18						
		0.25						0.05				
0.37		0.31	0.71									
0.43	0.84	0.88	1.00	0.95	0.78							
		0.93	1.00	1.00	1.00	0.73			0.09			
		0.67		0.99	0.99	0.79	0.48		0.10			
		0.91			0.77	0.63			0.09			
0.68	0.94	1.00	0.90									
0.78	1.00	0.68	0.92		0.42							
0.73	0.97	1.00	0.84							0.03		
0.55	0.72	0.75	0.65	0.44						0.03		

c.

			0.85	0.95	0.85							
0.01	0.96	1.00	1.00	1.00	0.96							
0.85	1.00	1.00	1.00	1.00	1.00	0.85						
0.95	1.00	1.00	1.57	1.00	1.00	0.95						
0.85	1.00	1.00	1.00	1.00	0.85							
0.96	1.00	1.00	1.00	0.96								
	0.85	0.95	0.85									

d.

0.32	0.42	0.45	0.41	0.30			0.30	0.41	0.45	0.42	0.32	
0.42	0.57	0.60	0.52	0.38			0.38	0.52	0.60	0.57	0.42	
0.45	0.59	0.61	0.50	0.04			0.04	0.50	0.61	0.59	0.45	
0.40	0.51	0.50	0.39				0.39	0.50	0.51	0.40		
0.30	0.35						0.35	0.30				
0.30	0.35						0.35	0.30				
0.40	0.51	0.50	0.39				0.39	0.50	0.51	0.40		
0.45	0.59	0.61	0.50	0.04			0.04	0.50	0.61	0.59	0.45	
0.42	0.57	0.60	0.52	0.38			0.36	0.52	0.60	0.57	0.42	
0.32	0.42	0.45	0.41	0.30			0.30	0.41	0.45	0.42	0.32	

FIGURE 13.6

Example equilibrium population arrangements in a 13×13 cell heterogeneous habitat patch with parameter settings $C = 36$, $r = .85$, $\mu = 2.5$, $b_{85} = 2.0$, and all other $b_i = 1.0$ given (a) no habitat preferences (i.e., lower bounds of zero for all v_i), (b) moderate preference for the center cell ($v_{85} \geq 1.0$), (c) strong preference for the center cell ($v_{85} \geq 1.5721$), and (d) strong preference for the four corner cells ($v_1, v_{13}, v_{157}, v_{169} \geq .3184$).

lower bounds identically for the populations in the four corner cells of the complex ($v_1, v_{13}, v_{157}, v_{169}$). At lower bound values of .3184 for each of the four cells, highly clustered equilibrium populations again result (figure 13.6d) with no further gain possible.

Discussion

Distribution and abundance patterns across a habitat complex are fundamentally influenced by how individuals within a population move and reproduce and by other factors limiting populations at various scales. We have represented population growth and movement with a simple reaction–diffusion formulation that models dispersal passively with distance decay across a lattice of breeding sites. Overlaying this reaction–diffusion process is a set of carrying capacity constraints at both the habitat cell scale (e.g., a breeding adult limit within each site) and the landscape scale (e.g., a predation effect limit). Mathematical programming was used to examine distribution and abundance patterns that achieve maximum equilibrium population conditions within a habitat patch subject to reaction–diffusion and carrying capacity constraints. The results from our experiments provide several new hypotheses regarding population distribution and abundance patterns.

In all our experiments, we observed that when a landscape-level carrying capacity constraint (equation 13.3) is strongly limiting and reaction–diffusion (equation 13.1) is less so, then many equally good (i.e., supporting the same total population) equilibrium population distributions with varying degrees of fragmentation are possible. This flexibility is inferred from the fact that we were able to demonstrate a broad range of distributions that showed both clustering and fragmentation (figures 13.1 and 13.2). Conversely, as reaction–diffusion processes become more limiting, possible population arrangements become more restrictive. Under strong reaction–diffusion limitations, little or no flexibility remains in how individuals can be arranged across the landscape, with all solutions converging toward contiguous Gaussian-like spatial distributions (subject to saturation capacity limits and preferences for individual habitat cells [figures 13.3, 13.5, 13.6c, and 13.6d]). In our model, reaction–diffusion processes become strongly limiting when reproduction and dispersal rates approach an extinction threshold (regardless of landscape-level limiting factors) and when site occupancy preferences are strong enough to cause reproduction and dispersal to become limiting, even though a small fraction of the population may actually use those sites.

Conventional wisdom might lead one to suspect that highly fragmented populations are subject to greater extinction risks because the organisms are only loosely connected in the habitat complex (Gilpin 1987). Although this may often be the case, our results indicate that some fragmented populations could instead be in a very viable state. In the presence of macroscale limiting factors, a fragmented population arrangement may be one of many alternative patterns that have essentially the same extinction risk. Similarly, when passively diffusing populations are highly clustered, population persistence may still be of concern, depending on how near the reaction–diffusion processes are to extinction threshold conditions.

An unanticipated result of our experiments is that equilibrium population clustering occurs when strong habitat preferences are introduced, even though few members of the clusters may actually occupy the preferred cells. Although we expected some local clustering within and around preferred cells, we did not expect to observe strong population attraction around preferred sites with passive diffusion. Population contiguity indices for the habitat preference experiments were among the highest we observed (table 13.1). We anticipated that such results would require biased diffusion models (e.g., Allen 1983) that include gregarious density-dependent movement or have a strong species association with an environmental variable that clusters spatially (Westman 1980). Instead we found that highly clustered populations can result from very localized site preference, as has been observed with species exhibiting strong site tenacity (Henschel and Lubin 1997). What appears to be gregarious behavior or a strong association between populations and some environmental gradient may in some cases stem from random diffusion processes in the presence of multiscale limiting factors and local site preferences.

Our results offer a potential explanation for why species-habitat association studies have been characterized by high variability (see Collins 1983). When occupancy patterns from similar sites are compared, variable results could arise from the existence of alternative equilibrium population arrangements even when those sites are subject to identical sets of limiting habitat factors. Furthermore, if population clusters occur across a landscape in response to local site preferences, then estimates of association between populations and environmental factors could be misleading or go undetected if conditions at the local site are not representative of conditions measured throughout the cluster (Orians and Wittenberger 1991).

These experiments suggest that interpreting the spatial arrangement of populations may be particularly difficult in the absence of fairly detailed information on population growth rates, dispersal, and the factors (including their scale of influence) limiting population size. However, our observations are tied to a specific model formulation, its attendant simplifications, and the limited parameter set explored in our numerical analyses. For example, we have examined the spatial pattern of populations within a closed system. We treated our landscape as a spatially autonomous patch with no immigration and with an absorbing boundary such that individuals dispersing beyond the patch were assumed to perish. Although this may be appropriate for modeling isolated systems, we would expect patterns within systems open to immigration to exhibit smaller habitat size thresholds (see Pagel and Payne 1996) resulting in flexibility for equilibrium population arrangements over a broader range of reproduction and dispersal parameter values. Likewise, we have examined only a single-patch system, but critical habitat size and arrangement thresholds have also been shown to exist for multipatch reaction–diffusion systems (chapter 6). In such systems, we would expect any increases in disperser survivorship resulting from biased diffusion or a somewhat hospitable interpatch matrix also to extend the range of parameter values over which many equally good population arrangements could be observed.

Movements underlying the ecological process represented by the land-scape-level constraint (equation 13.3) are also an important consideration in the applicability of our model. At the finer scale, organism movements are modeled by a reaction–diffusion process. When the landscape-level constraint represents activities by other system components such as predators or distur-bance agents, our formulation seems reasonable. On the other hand, when the landscape-level constraint represents processes involving movements by the same organisms as those modeled at the finer scale, as in a forage constraint (see chapter 10), this formulation may apply only in fairly limited cases. It may be appropriate for species that have a strong fidelity to natal breeding sites and that obtain much of their food supply from the surrounding land-scape, but we do not expect this formulation to be applicable to territorial species in general. Energetic costs (Covich 1976; Orians and Pearson 1979) or sensitivity to intraspecific interference (Dolman and Sutherland 1997) associated with foraging may limit food acquisition to areas that are in prox-imity to breeding sites. The effects of restricting foraging movements to account for energetic costs were examined in chapter 10. That formulation resulted in smaller breeding clusters that were more dispersed throughout the landscape than we observed in this chapter. Although the species distribu-tional patterns observed with a foraging movement limitation in chapter 10 differed from those we report here, they are not contradictory results. Rather, they tend to emphasize the point that fragmented populations may be very viable as an outcome of multiscale limitations.

Appendix

The degree of clustering in population distributions is estimated by first con-verting each solution to a binary map (if $v_i > 0$, then $v_i = 1$, otherwise $v_i = 0$) reflecting the pattern of cell occupancy across the landscape. Clustering in binary maps can be tested using join-count statistics (Dacey 1968; Cliff and Ord 1981) where two cells sharing a boundary within a lattice are said to be linked by a join. Clustering in the pattern of cell occupancy is determined by counting the number of joins between cells that are first-order neighbors (cells that share a common edge or vertex) and comparing it with the number expected under the null hypothesis of no spatial autocorrelation between the cells. Following the procedures in Gatrell (1977:36), we estimate z_{mn} as a measure of the deviation between observed join-counts and the count expected under the null hypothesis of spatial independence between map elements m and n, where m and n take on values of zero for the unoccupied map element and 1 for the occupied map element in our case. A spatial clustering statistic is estimated as

$$z_{mn} = \frac{O_{mn} - E_{mn}}{\sigma_{mn}},$$

where O_{mn} is the observed join-count between map elements m and n, E_{mn} is the expected count under the null hypothesis of no spatial dependence, σ_{mn} is the standard deviation, and z_{mn} is interpreted as a standard normal deviate. Formulas for E_{mn} and σ_{mn} under sampling without replacement are given by Cliff and Ord (1981:20). We are interested in the pattern among occupied cells. Therefore $z_{11} \gg 0$ indicates a tendency for occupied cells to be clustered, and $z_{11} \ll 0$ indicates hyperdispersion or avoidance among occupied cells.

Concentration of population abundance is estimated as

$$K = \frac{\sum_{i=1}^{N} v_i / b_i}{\sum_{i=1}^{N} \delta_i},$$

where $\delta_i = 1$ when $v_i > 0$. Therefore, K is simply the mean standardized abundance observed in occupied cells. Because we have standardized abundance as a simple index reflecting the proportion of occupied cellular carrying capacity, our statistic for abundance approaches 1.0 under maximum concentration.

14

CARBON FIXATION IN TREES
AS AN OPTIMIZATION PROCESS

In chapter 13 we used optimization as a device for finding model equilibria and then used that model to develop hypotheses about ecosystem function. In this chapter we directly apply optimization as a simulation of ecological optimizing behavior. To this end, we investigate economic analogues to ecological behavior. Applying economic models to ecological problems is not a new concept. For example, Rapport and Turner (1977:367) wrote,

> Ecological processes have traditionally been studied from several vantage points. . . . None of these approaches, however, explicitly address what some . . . have regarded as one of the central problems of ecology— the ways in which scarce resources are allocated among alternative uses and users. The question is, of course, fundamental to economic thinking . . . and it is for this reason that we have recently seen the introduction of essentially economic models and modes of thought in ecology.

Progress in this direction has been slow. Bazzaz and Grace (1997) survey a large body of work on plant resource allocation, for example, and summarize a considerable amount of literature that has applied economic or optimization methods to the topic. Specific model formulations and empirical applications still appear to be rare, however, and their excellent book calls for much more work in this area. Bloom et al.'s (1985:363) comment that "Plant ecologists . . . have

---

This chapter was adapted from J. Hof, D. Rideout, and D. Binkley, Carbon fixation in trees as a micro optimization process: An example of combining ecology and economics, *Ecological Economics* 2 (1990): 243–256, with permission from the publisher, Elsevier Science.

yet to evaluate fully the applicability of economic theory to plant processes" still appears to be essentially true. This chapter investigates a modification of traditional economic models of optimizing behavior that may be applicable to a variety of ecological systems, using carbon fixation in trees as an illustration.

Formulation

In Hof et al. (1990), we assume that trees behave in a manner that maximizes net carbon gain during a growing season. Rationales for this assumption are readily available in the biological literature (e.g., Bloom et al. 1985; Chapin et al. 1987; Fitter and Hay 1987). Most forest productivity models have focused on the production of biomass given environmental constraints (cf. Graham et al. 1985; Byrne et al. 1986; Running and Coughlan 1988) and have not addressed the challenging question of predicting the optimal allocation of plant components to obtain the resources necessary for maximizing the overall net gain of carbon. Bloom et al. (1985) developed some predictions about plant productivity from a verbal model based on some simple economic concepts. These authors emphasize the importance of flexibility in carbohydrate allocation between leaves and roots to obtain the optimum mix of resources for overall plant carbohydrate gain. Maximizing net carbon gain is a plausible plant objective because carbon enables plants to grow, which in turn allows them to compete with other plants for light and water by building leaves and roots. Bloom et al. conclude that optimal carbohydrate allocation in plants would exist when all resources (such as water, light, and nutrients) were equally limiting to plant production. They expressed this condition in terms of exchange ratios (the cost of obtaining a resource that will provide a unit of carbohydrate gain) and argue that optimal allocation in plants leads to equal exchange ratios between resources. For example, a plant stressed by low water availability would balance its allocation between leaves and roots such that an additional unit of carbon expenditure on more leaves or more roots would provide the same carbohydrate gain.

Carbon fixation requires both water and energy (light) in addition to carbon dioxide as a carbon source. The tree must expend carbon on roots and leaves to obtain this water and light energy. We do not consider the possible responses of plants to changes in carbon dioxide concentration in the atmosphere or to nutrient limitations. A model of optimized leaf and root construction and maintenance that is an analog of traditional profit maximization is as follows:

Maximize

$$N = C[E(L), H(R)] - W_L L - W_R R,$$

where N is net carbon gain, C is gross carbon gain, E is light energy obtained from leaf area, L is leaf area, H is water obtained from root biomass, R is root biomass, W_L is unit carbon cost of producing and maintaining leaf area, and W_R is unit carbon cost of producing and maintaining root biomass.

As an analogue to a profit maximization model, this model treats gross carbon gain as "revenue," light energy and water obtained by the plants as "outputs," and leaves and roots as "inputs" to producing those outputs. This model reflects a multistage production process in which carbon is expended to produce leaves and roots (at unit costs W_L and W_R), which are then used to obtain light energy and water, which in turn are used to fix carbon. The light energy and water are extracted from the environment and are inputs to carbon fixation, but the light and water extracted are outputs to the plant in the same sense that precious metals extracted from a mine are outputs to the mining company. As with a mine, the rates of extraction for the plant are affected by the concentration of the (light and water) outputs in the environment. Because of the relationships between leaves and light energy and between roots and water, leaf area and root biomass are the only choice variables. Note that the production functions $E(L)$ and $H(R)$ are potentially nonlinear, as may be the carbon "revenue" function.

The first-order conditions for this model are

$$\frac{\partial N}{\partial L} = \frac{\partial C}{\partial E}\frac{\partial E}{\partial L} - W_L = 0,$$

$$\frac{\partial N}{\partial R} = \frac{\partial C}{\partial H}\frac{\partial H}{\partial R} - W_R = 0,$$

or, more compactly,

$$\frac{\partial C}{\partial L} = W_L,$$

$$\frac{\partial C}{\partial R} = W_R,$$

which implies that

$$\frac{W_L}{\partial C/\partial L} = \frac{W_R}{\partial C/\partial R} = 1. \tag{14.1}$$

This condition is consistent with that hypothesized by Bloom et al. (1985:367): "By this theorem, a plant should adjust allocation so that, for a given expenditure in acquiring each resource, it achieves the same growth response: Growth is equally limited by all resources." Bloom et al. assume that plants maximize growth rather than carbon gain. They provide no test of this theorem.

This model has much appeal, but the theme of this chapter is that such a model may not adequately capture the way in which many ecological systems behave. Let us examine the process of the carbon fixation example further. The carbohydrate gained per unit of light intercepted by the leaves is limited by the rate of carbon dioxide uptake through the leaf stomata; stomatal conductance is regulated largely by the supply of water available for transpiration. Thus the carbon gain might potentially be limited by either the leaf area that collects carbon dioxide and light energy or by the root area that collects water, which in turn enables the stomata to capture the carbon dioxide.

We thus hypothesize that the plant will attempt to maximize the minimum of two functions: the net carbon gain achieved from leaf area assuming adequate water is available and the net carbon gain achieved from room biomass assuming adequate carbon dioxide and light energy are available. We are *not* hypothesizing that plant carbon fixation is limited by a single factor. As will be seen, the optimal solutions to this model imply that carbon is limited by both leaves and roots, but not as implied by the previous model (condition 14.1). Accounting for leaves or roots being *potentially* nonlimiting may be a more realistic model of tree behavior.

More formally, the hypothesized model is as follows:

Choose L, R, and A so as to maximize

$$N = A - W_L L - W_R R,$$

subject to

$$A - C_L (L) \leq K_L = 0,$$
$$A - C_R (R) \leq K_R = 0,$$

where N is net carbon gain, A is the minimum of C_L and C_R and is gross carbon gain, C_L is the gross carbon gain from leaf area if adequate water is available, C_R is the gross carbon gain from root biomass if adequate light and carbon dioxide are available, and the Ks are right-hand sides set to zero.

The Lagrangean function (M), with multipliers λ_L and λ_R, is

$$M = A - W_L L - W_R R + \lambda_L [K_L - (A - C_L(L))] + \lambda_R [K_R - (A - C_R(R))],$$

or

$$M = A - W_L L - W_R R + \lambda_L[C_L(L) - A] + \lambda_R[C_R(R) - A].$$

The Karush–Kuhn–Tucker conditions for a saddle point are

$$\frac{\partial M}{\partial L} \leq 0 \quad L \geq 0 \quad \text{and} \quad L\frac{\partial M}{\partial L} = 0,$$

$$\frac{\partial M}{\partial R} \leq 0 \quad R \geq 0 \quad \text{and} \quad R\frac{\partial M}{\partial R} = 0,$$

$$\frac{\partial M}{\partial A} \leq 0 \quad A \geq 0 \quad \text{and} \quad A\frac{\partial M}{\partial A} = 0,$$

$$\frac{\partial M}{\partial \lambda_L} \geq 0 \quad \lambda_L \geq 0 \quad \text{and} \quad \lambda_L\frac{\partial M}{\partial \lambda_L} = 0,$$

$$\frac{\partial M}{\partial \lambda_R} \geq 0 \quad \lambda_R \geq 0 \quad \text{and} \quad \lambda_R\frac{\partial M}{\partial \lambda_R} = 0.$$

If C_L and C_R are strictly concave (assumed here), these conditions are necessary and sufficient for an optimum.

The case of interest is an interior solution ($L > 0, R > 0, A > 0$) where both constraints are binding ($\lambda_L > 0, \lambda_R > 0$). Clearly, trees are expected to behave such that $L > 0, R > 0$, and $A > 0$, and multiple limiting factors are anticipated as long as W_L and W_R are nonzero (positive). In this case, the Karush–Kuhn–Tucker conditions reduce to

$$\frac{\partial M}{\partial L} = 0 \quad \frac{\partial M}{\partial R} = 0 \quad \frac{\partial M}{\partial A} = 0 \quad \frac{\partial M}{\partial \lambda_L} = 0 \quad \frac{\partial M}{\partial \lambda_R} = 0,$$

which indicates that

$$\lambda_L \frac{\partial C_L}{\partial L} = W_L,$$

$$\lambda_R \frac{\partial C_R}{\partial R} = W_R,$$

$$\lambda_L + \lambda_R = 1,$$

$$C_L(L) = A,$$

$$C_R(R) = A.$$

The Karush–Kuhn–Tucker conditions thus imply that

$$\frac{W_L}{\partial C_L/\partial L} + \frac{W_R}{\partial C_R/\partial R} = 1, \tag{14.2}$$

which is in sharp contrast to condition (14.1). Condition (14.2) is consistent with Bloom et al. (1985) in that L and R simultaneously limit carbon gain, but it differs in that they are not necessarily equally limiting. That is, the exchange ratios for water and light obtained per unit of carbon expended on roots and leaves are not equal to one another but sum to 1. The MAXMIN approach indicates that optimal plants may be simultaneously limited by more than one resource, but all exchange ratios are not equal.

Denoting optimized values with asterisks,

$$\lambda_L^* = \lambda_L = \frac{\partial M}{\partial K_L} = \frac{\partial N^*}{\partial K_L},$$

$$\lambda_R^* = \lambda_R = \frac{\partial M}{\partial K_R} = \frac{\partial N^*}{\partial K_R},$$

at the optimal solution point, by the envelope theorem. Thus at the solution point, the Karush–Kuhn–Tucker conditions indicate that

$$\frac{\partial N^*}{\partial K_L} = \frac{\partial C_L^*}{\partial L^*} = W_L,$$

$$\frac{\partial N^*}{\partial K_R} = \frac{\partial C_R^*}{\partial R^*} = W_R,$$

which specifies that L (R) will be used until its marginal product in producing C_L (C_R) times the shadow price of the first (second) constraint equals W_L (W_R).

Conditions (14.1) and (14.2) provide an observable, testable hypothesis for supporting or rejecting the more traditional model as opposed to the MAXMIN model just developed. This provides an example of how optimization procedures common in microeconomic theory might be used to direct biological research in developing testable hypotheses regarding ecological behavior. Further insights into the nature of the MAXMIN model are obtained in this chapter with an empirical demonstration of the model, using representative data. An advantage of the MAXMIN model is that the C_L and C_R functions appear to be more easily empiricized from extant research than the single C function in the first model. However, the mutually exclusive hypotheses represented by conditions (14.1) and (14.2) could be tested without either model being parameterized.

Case Example

To explore and demonstrate the MAXMIN model, we built an empirical example with information from the literature. The interception of light by leaves is determined by the Lambert–Beer equation:

Energy absorbed (MJ/m^2) = Incoming radiation $\times$ $(1 - \exp(K\ LAI))$,

where K is the light extinction coefficient and LAI is the leaf area index of the plant or stand. We used $-.45$ for K and converted leaf biomass into LAI assuming 1 kg of leaves provided 4 m^2 of projected leaf area (similar to conifer leaves). The carbohydrate gain per unit of light energy absorbed was assumed to be .003 kg CH_2O/MJ (.0014 kg C/MJ) of energy (if water were available), based on Jarvis and Leverentz (1983) and Jarvis (1985). Respiration rates are not well known for trees. Our best estimate of the cost of synthesizing roots and leaves was 1.3 kg CH_2O/kg of tissue produced (representing 1 kg of new tissue and .3 kg of respired CH_2O; Penning de Vries 1983). The cost of maintaining leaves for 1 year was assumed to be .5 kg CH_2O/kg, and the cost of root maintenance was assumed to be .25 kg CH_2O/kg.

The carbohydrate gained per unit of light intercepted is limited by the rate of CO_2 uptake through stomata, and stomatal conductance is regulated in part by the water available for transpiration. The carbohydrate gain per unit of light intercepted usually appears to be linear across a gradient of water supply, with the possible exception of severe drought conditions (Jarvis 1985). Therefore, we used a constant gain of .075 kg CH_2O/m^2 for each centimeter of water obtained by the roots. This value is a representative average based on simulations of forest productivity from a wide range of sites in North America by Running and Coughlan (1988).

We found no experimental information on the relationship between root biomass and water uptake, so we assumed that water uptake followed Michelis–Menten mass action kinetics:

$$\text{Water uptake} = \frac{\text{Soil water (cm)} \times \text{Root biomass}}{.1 + \text{Root biomass}},$$

where .1 represents the root biomass (in kilograms per square meter) needed to obtain half of the water available for uptake.

These relationships were combined to produce a MAXMIN model as follows:

Maximize

$$N = A - (1.3 + .5)L - (1.3 + .25)R, \tag{14.3}$$

subject to

$$(.0014E_I)\{1 - \exp[-(.45 \times 4)L]\} = C_L, \tag{14.4}$$

$$(.075S_H)\frac{R}{.1 + R} = C_R, \tag{14.5}$$

$$A - C_L \leq K_L = 0, \tag{14.6}$$

$$A - C_R \leq K_R = 0, \tag{14.7}$$

$$\text{LAI} = 4L, \tag{14.8}$$

where N, A, L, R, C_L, and C_R are as defined before, S_H is soil water (40 cm), E_I is incoming light (4000 MJ/m^2), and LAI is leaf area index.

Results

Basic Solution

We solved the model using a version of the generalized reduced gradient algorithm for solving nonlinear programs (originally developed by Abadie 1978; see also Liebman et al. 1986). Tolerances were set such that feasibility is ensured within 1×10^{-6}. No advanced starts were needed, and first-order conditions were met within 1×10^{-3} for all solutions presented.

The solution to this model is shown in table 14.1. From that solution,

$$\frac{\partial C_L}{\partial L} = -5.6\,(\exp(-1.8L) - 1.8) = 6.13994 \quad \text{when } L = .27541,$$

$$\frac{\partial C_R}{\partial R} = \frac{.3}{(.1 + R)^2} = 2.19280 \qquad \text{when } R = .26988.$$

From the shadow prices on (14.6) and (14.7),

$$\lambda_L = .293165,$$
$$\lambda_R = .706872.$$

With these results, all of the Karush–Kuhn–Tucker conditions can be verified:

$$\lambda_L \frac{\partial C_L}{\partial L} = 1.800015 \approx W_L,$$

$$\lambda_R \frac{\partial C_R}{\partial R} = 1.550029 \approx W_R,$$

$$\lambda_L + \lambda_R = 1.000037 \approx 1.$$

TABLE 14.1
Solution to MAXMIN Model

Variable	Value	Reduced Cost
A	2.188923	.000037
L	.275410	.000000
R	.269879	.000000
C_L	2.188923	.000000
C_R	2.188923	.000000
LAI	1.101641	.000000

Constraint	Slack or Surplus	Price
14.4	.000000	−.293165
14.5	.000000	−.706872
14.6	.000000	.293165
14.7	.000000	.706872
14.8	.000000	.000000

Objective function value, 1.274873.

Both constraints in (14.6) and (14.7) are binding, and $A > 0$, $L > 0$, $R > 0$ (an interior solution was achieved).

It should also be pointed out that

$$\frac{W_L}{\partial C_L / \partial L} = .293165, \qquad \frac{W_R}{\partial C_R / \partial R} = .706872,$$

indicating that, in solution, condition (14.2) is sharply different from condition (14.1). If the MAXMIN model is a good representation of tree carbon fixation, the more traditional model probably is a poor one.

Sensitivity Analysis

To explore the behavior of the MAXMIN model, we repeatedly solved it with varying parameter levels. Reducing the light interception by raising the light extinction coefficient (K) from −.45 to −.30 decreased net carbohydrate gain by 19%, and increasing interception by lowering K to −.60 increased net carbohydrate gain by 10%. Lowering the efficiency of converting intercepted light into carbohydrates from .003 to .0015 kg CH_2O/MJ reduced net carbohydrate gain by about 16%. Increasing the respiration cost of synthesizing new tissues from 1.3 to 1.6 kg CH_2O/kg of new tissue lowered net carbohydrate gain by 12%. Increasing the maintenance cost of roots from .25 to .5 CH_2O/kg decreased net carbohydrate gain by 5%, and increasing the maintenance cost of leaves from .5 to 1.0 kg CH_2O/kg decreased net carbohydrate gain by 10%. However, the optimal allocation pattern between roots and leaves was insensitive to variations in respiration parameters relative to changes in other parameters.

The net gain of carbohydrate was fairly sensitive to water use efficiency. Increasing the efficiency of carbohydrate capture per unit water from .075 to .10 kg CH_2O/cm H_2O increased net carbohydrate gain by 34%, and decreasing the efficiency to .050 kg CH_2O/cm H_2O decreased net carbohydrate gain by 42%. Leaf area index was much more sensitive to changes in water use efficiency than was root biomass.

At low water availability (< 40 cm), the optimal biomass of leaves declined with increasing light. When water was more available, optimal leaf biomass increased with increasing light up to about 3500 MJ/m^2 and then declined with increasing light. At low light (below about 3000 MJ/m^2), the optimal root biomass increased as the availability of water decreased. At intermediate levels of light (3000–5000 MJ/m^2), optimal root biomass was highest at intermediate water availabilities (about 40–60 cm), decreasing at higher and lower water levels. The optimal root biomass consistently increased with increasing light. The optimal ratio of leaf to root biomass consistently increased with increasing water availability and with decreasing light.

Our model always produced optimal solutions that resulted in roots and leaves simultaneously limiting carbon gain. The leaf biomass did not grow large enough to intercept any more light than could be converted into carbohydrate with the amount of water available from the roots. Similarly, root biomass was not increased beyond the minimum necessary to provide the quantity of water needed to produce the maximum quantity of carbohydrates allowed by the quantity of light intercepted. It is not surprising that the pattern of optimal allocation led to simultaneous limitations of root and leaf biomass (Chapin et al. 1987). Perhaps less obvious is that this resulted in the quantity of carbohydrate produced per unit of light intercepted or per unit of water obtained being constant across gradients in the availability of light and water. In fact, the gross carbohydrate gain per unit of light intercepted was .003 kg CH_2O/MJ (.0015 kg C/MJ), which is simply the conversion efficiency we used from Jarvis and Leverentz (1983) and Jarvis (1985). Similarly, the gross carbohydrate gain per unit of water used was the conversion efficiency .075 kg/cm of water, which is the conversion efficiency we specified from Running and Coughlan (1988). Therefore the variations in gross carbon gain with different light and water availabilities resulted from differences in the quantities of light intercepted or water obtained rather than from variations in the efficiencies with which each resource was used. This observation is analogous to a constant market price per unit of product that is independent of the quantity of product sold.

Discussion

The implications of our model for the efficiency of net carbohydrate production per unit of light intercepted or water obtained may be unexpected. We found that for any given availability of water, the efficiency of net carbohydrate produced per unit of light intercepted increased. Fixed water availability sets a limit on the possible carbon gained from light interception, and increases in the availability of light allow the same quantity of light to be intercepted by fewer leaves. Therefore the carbon cost (of leaves) declines per unit of light intercepted as the availability of light increases. For analogous reasons, the efficiency of net carbohydrate production per unit of water obtained increased as the availability of water increased for any given availability of light. Bloom et al. (1985) define biomass (or grams of C) as representing both products and revenue and water as a resource (or input), and by these definitions our model demonstration appears incongruous. However, in our analogy we define inputs as the carbon invested in leaves and roots, products are the captured light and water, and revenues are the gross and net carbon gains. This illustrates the importance of keeping the definitions of economic and ecological analogues clear. Our demonstration of increasing net carbon gain as the availability of light and water increased is analogous to increasing profit per unit of product caused by a decline in input cost per unit of product produced (each leaf intercepted more light, or each root obtained more water). Again, this resembles the economics of a mining operation, where the profit per unit of metal extracted increases as the availability (or concentration) of the metal increases in the ore that is processed.

These outcomes of the MAXMIN approach may not be met by plants growing under field conditions because of the time frame involved for producing roots and leaves relative to rates of change in the availability of light and water. The availability of light and water tends to vary over a smaller time scale than the time needed to adjust the biomass of roots and leaves. Plants face an especially interesting challenge to optimize the production of roots and leaves in an environment where the value of roots and leaves (in terms of water and light captured) varies more quickly than the response of the plants. This challenge is also common for firms that operate in dynamic markets.

Both the theoretical model and the empirical demonstration provide examples of how optimization might be used to guide ecological research in directions that imply tests of fundamental behavioral postulates. Given the nearly infinite number of potential ecological tests, hypothesis development such as that demonstrated here might be a useful approach to improving our understanding of ecological systems.

This completes our fourth and final section of applications. Chapter 15 offers a brief postscript to conclude the book.

15

POSTSCRIPT

The principal thesis of this book has been that readily solvable mathematical programming models can play an important role in spatial modeling for landscape management and ecological research. Numerous spatial optimization examples were presented in the first three parts, and two theoretical ecology examples were presented in part IV. Linear programming models that can be solved for large-scale spatial problems were used in nearly all these examples.

Looking across the linear programming models that have been presented here, we are able to observe several solution properties (in addition to general spatial sensitivity) that appear to capture well-recognized hydrologic and ecological characteristics. It is encouraging that we obtain these effects even from models with such simple assumptions as constant periodic reproduction and dispersal. Some examples of well-recognized effects resulting from linear programming models in our chapters include the following:

Realistic stormflow hydrographs	Chapter 3
Sigmoid population growth curves	Chapter 7
Propagation waves of invasion	Chapters 7 and 11
Population clustering	Chapters 7 and 13
Saturation-dependent dispersal	Chapter 8
Sensitivity to climatic variation	Chapter 9
Negative and positive edge effects	Chapter 10
Habitat extinction thresholds	Chapter 10
Quasirandom distributions	Chapter 13

Linear programming models can also be useful for approximating results from the more detailed dynamics often included in simulation models. In chapter 5, for example, we showed that spatially defined first-order relationships estimated from a complex simulation model for the northern spotted owl (McKelvey et al. 1992) could be used to construct a linear programming model that can approximate and improve upon simulation results (Hof and Raphael 1997).

These results are encouraging considering the range of realistic ecological effects that can be portrayed, but we do not mean to imply that linear programming models are the answer to all problems. Rather, our interpretation is that in the interests of parsimony and solvability, linear models should be of some (and perhaps substantial) use in ecology, particularly when one considers that no model provides the definitive answer to real-world problems. Given past misconceptions that linear programming models are unsuitable for handling spatial problems (e.g., Bettinger et al. 1996; Van Deusen 1996; Murray and Snyder 2000) or problems such as those involving augmentation of habitat in an initially fragmented system (Loehle 1999), perhaps our most important message has been to show otherwise.

Integer and nonlinear optimization models can also be readily solvable, as chapters 4 and 14 respectively demonstrate, but this is often not the case. As Boston (1999) suggests, heuristic solution procedures (e.g., Reeves 1993) may offer a reasonable approach for solving large-scale nonlinear and integer problems, although much research along those lines remains to be done (Jager and Gross 2000). Problems that combine both integer and nonlinear requirements or that have a high degree of nonconvexity can still be very challenging.

A number of other important areas for further research on the subject of spatial optimization modeling remain. Four research areas seem particularly critical, in addition to the obvious need for further case studies and empirical tests. First is the continuing quest to capture greater ecological detail (that may necessitate the integer and nonlinear models just discussed) and to solve the numerical problems this implies. Jager and Gross (2000) emphasize this challenge in their review of our previous book. Second is to improve our understanding of and accounting for stochastic effects. Discrete-event Monte Carlo simulation models are widely used to examine the effects of random demographic and environmental variability in complex systems, along with more theoretical treatments of stochastic processes. However, little of this work focuses on optimization, and many research opportunities remain. A third area for further research involves expanding this work to develop community-level models. To sustain viable populations in managed (or at least affected) systems, we will need to better understand how to model the community interaction effects of our treatment activities. A fourth important research area is monitoring. If models such as ours are to be used in an adaptive management process, monitoring methods will be needed to interpret real-world effects and suggest model modifications. With more understanding in each of these areas, we should be positioned to develop better process-oriented spatial optimization models to help design and implement management strategies for the complex ecosystems we inhabit.

REFERENCES

Abadie, J. 1978. The GRG method for nonlinear programming. In H. J. Greenberg, ed., *Design and Implementation of Optimization Software,* pp. 335–367. Amsterdam: Sijthoff-Noordhoff/Kluwer Academic.

Adams, D. M., and A. R. Ek. 1974. Optimizing the management of uneven-age forest stands. *Canadian Journal of Forest Research* 4:274–287.

Adler, F. R., and B. Nuernberger. 1994. Persistence in patchy irregular landscapes. *Theoretical Population Biology* 45:41–75.

Allen, L. J. S. 1983. Persistence and extinction in single-species reaction–diffusion models. *Bulletin of Mathematical Biology* 45:209–227.

Allen, L. J. S. 1987. Persistence, extinction, and critical patch number for island populations. *Journal of Mathematical Biology* 24:617–625.

Ammerman, A. J., and L. L. Cavalli-Sforza. 1984. *The Neolithic Transition and the Genetics of Populations in Europe.* Princeton, N.J.: Princeton University Press.

Anderson, D. H. 1989. A mathematical model for fire containment. *Canadian Journal of Forest Research* 19:997–1003.

Anderson, D. J., and B. B. Bare. 1994. A dynamic programming algorithm for optimization of uneven-aged forest stands. *Canadian Journal of Forest Research* 24:1758–1765.

Anderson, E., S. C. Forrest, T. W. Clark, and L. Richardson. 1986. Paleobiology, biogeography, and systematics of the black-footed ferret, *Mustela nigripes* (Audubon and Bachman) 1851. *Great Basin Naturalist Memoirs* 8:11–62.

Andow, D. A., P. M. Kareiva, S. A. Levin, and A. Okubo. 1990. Spread of invading organisms. *Landscape Ecology* 4:177–188.

Andrews, P. L. 1986. *BEHAVE: Fire Behavior Prediction and Fuel Modeling System— BURN Subsystem, part 1,* General Technical Report INT-194. Ogden, Utah: USDA Forest Service.

Aneja, Y. P., and M. Parlar. 1984. Optimal staffing of a forest fire fighting organization. *Canadian Journal Forest Research* 14:589–594.

Anonymous. 1998a. MacMillan Bloedel to stop clearcutting. *Forestry Source* 3(7):1, 13.

Anonymous. 1998b. Scientists endorse controversial forest management bill. *Forestry Source* 3(6):1, 1b.

Apa, A. D., D. W. Uresk, and R. L. Linder. 1990. Black-tailed prairie dog populations one year after treatment with rodenticides. *Great Basin Naturalist* 50:107–113.

Arno, S. F., and J. K. Brown. 1991. Overcoming the paradox in managing wildland fire. *Western Wildlands* 17:40–46.

Bare, B. B., and D. Opalach. 1988. Determining investment-efficient diameter distributions for uneven-aged northern hardwoods. *Forest Science* 34:243–249.

Barnes, A. M. 1993. A review of plague and its relevance to prairie dog populations and the black-footed ferret. In J. L. Oldemeyer, D. E. Biggins, B. J. Miller, and R. Crete, eds., *Proceedings of the Symposium on the Management of Prairie Dog Complexes for the Reintroduction of Black-Footed Ferrets,* Biological Report No. 13, pp. 28–37. Washington, D.C.: U.S. Fish and Wildlife Service.

Bart, J. 1995. Amount of suitable habitat and viability of northern spotted owls. *Conservation Biology* 9:943–946.

Bart, J., and E. D. Forsman. 1992. Dependence of northern spotted owls *Strix occidentalis caurina* on old-growth forest in the western USA. *Biological Conservation* 62:95–100.

Bascompte, J., and R. V. Solé. 1994. Spatially induced bifurcations in single-species population dynamics. *Journal of Animal Ecology* 63:256–264.

Batabyal, A. A. 1998. An optimal stopping approach to the conservation of biodiversity. *Ecological Modelling* 105:293–298.

Bazzaz, F. A., and J. Grace. 1997. *Plant Resource Allocation.* San Diego, Calif.: Academic Press.

Bettinger, P., K. N. Johnson, and J. Sessions. 1996. Forest planning in an Oregon case study: Defining the problem and attempting to meet goals with a spatial-analysis technique. *Environmental Management* 20:565–577.

Bettinger, P., J. Sessions, and K. Boston. 1997. Using Tabu search to schedule timber harvests subject to spatial wildlife goals for big game. *Ecological Modelling* 42:111–123.

Bettinger, P., J. Sessions, and K. N. Johnson. 1998. Ensuring the compatibility of aquatic habitat and commodity production goals in eastern Oregon with a tabu search procedure. *Forest Science* 44:96–112.

Bevers, M., and C. H. Flather. 1999a. The distribution and abundance of populations limited at multiple spatial scales. *Journal of Animal Ecology* 68:976–987.

Bevers, M., and C. H. Flather. 1999b. Numerically exploring habitat fragmentation effects on populations using cell-based coupled map lattices. *Theoretical Population Biology* 55:61–76.

Bevers, M., J. Hof, B. Kent, and M. G. Raphael. 1995. Sustainable forest management for optimizing multispecies wildlife habitat: A coastal Douglas-fir example. *Natural Resource Modeling* 9:1–23.

Bevers, M., J. Hof, and C. Troendle. 1996. Spatially optimizing forest management schedules to meet stormflow constraints. *Water Resources Bulletin* 32:1007–1015.

Bevers, M., J. Hof, D. W. Uresk, and G. L. Schenbeck. 1997. Spatial optimization of prairie dog colonies for black-footed ferret recovery. *Operations Research* 45: 495–507.

Biggins, D. E., B. J. Miller, L. R. Hanebury, B. Oakleaf, A. H. Farmer, R. Crete, and A. Dood. 1993. A technique for evaluating black-footed ferret habitat. *U.S. Fish and Wildlife Service Biological Report* 93(13):73–88.

Biggins, D. E., M. H. Schroeder, S. C. Forrest, and L. Richardson. 1986. Activity of radio-tagged black-footed ferrets. *Great Basin Naturalist Memoirs* 8:135–140.

Bloom, A. J., F. S. Chapin, and H. A. Mooney. 1985. Resource limitation in plants: An economic analogy. *Annual Review of Ecology and Systematics* 16:363–392.

Boston, K. 1999. Review: *Spatial Optimization for Managed Ecosystems. Forest Science* 45:595.

Bottoms, K. E., and E. T. Bartlett. 1975. Resource allocation through goal programming. *Journal of Range Management* 28:442–447.

Bowles, M. L. 1983. The tallgrass prairie orchids *Platanthera leucophaea* (Nutt.) Lindl. and *Cypripedium candidum* Muhl. ex Willd.: Some aspects of their status, biology, and ecology, and implications toward management. *Natural Areas Journal* 3:14–37.

Bowles, M., R. Flakne, and R. Dombeck. 1992. Status and population fluctuations of the eastern prairie fringed orchid in Illinois. *Erigenia* 12:26– 40.

Brooke, A., D. Kendrick, and A. Meeraus. 1992. *GAMS: A User's Guide*. Danvers, Mass.: The Scientific Press Series.

Brown, A. A., and K. P. Davis. 1973. *Forest Fire Control and Use,* 2nd ed. New York: McGraw-Hill.

Buongiorno, J., and B. R. Michie. 1980. A matrix model of uneven-aged forest management. *Forest Science* 26:609–625.

Byrne, G., J. Landsberg, and M. Benson. 1986. The relationship of above-ground dry matter accumulation by *Pinus radiata* to intercepted solar radiation and soil water status. *Agricultural and Forest Meteorology* 37:63– 73.

Çambel, A. B. 1993. *Applied Chaos Theory: A Paradigm for Complexity.* San Diego, Calif.: Academic Press.

Chapin, F. S., A. J. Bloom, C. B. Field, and R. H. Waring. 1987. Plant responses to multiple environmental factors. *BioScience* 37:49–57.

Chiang, A. C. 1974. *Fundamental Methods of Mathematical Economics,* 2nd ed. New York: McGraw-Hill.

Cincotta, R. P. 1985. *Habitat and Dispersal of Black-Tailed Prairie Dogs in the Badlands National Park.* M.S. thesis, Colorado State University, Fort Collins.

Cincotta, R. P., D. W. Uresk, and R. M. Hansen. 1987. Demography of black-tailed prairie dog populations re-occupying sites treated with rodenticide. *Great Basin Naturalist* 47:339–343.

Cincotta, R. P., D. W. Uresk, and R. M. Hansen. 1988. A statistical model of expansion in a colony of black-tailed prairie dogs. In D. W. Uresk, G. L. Schenbeck, and R. Cefkin, tech. coords., *Eighth Great Plains Wildlife Damage Control Workshop Proceedings,* General Technical Report RM-154, pp. 30–33. Rapid City, S.Dak.: USDA Forest Service.

Clark, T. W. 1989. *Conservation Biology of the Black-Footed Ferret,* Mustela nigripes, Special Scientific Report No. 3. Philadelphia: Wildlife Preservation Trust International.

Cliff, A. D., and J. K. Ord. 1981. *Spatial Processes: Models and Applications.* London: Pion Limited.

Collins, S. L. 1983. Geographic variation in habitat structure for the wood warblers in Maine and Minnesota. *Oecologia* 59:246–252.

Conlin, W. M., and R. H. Giles, Jr. 1973. Maximizing edge and coverts for quail and small game. In *Proceedings of the First National Bobwhite Quail Symposium.* Stillwater: Oklahoma State University.

Covich, A. P. 1976. Analyzing shapes of foraging areas: Some ecological and economic theories. *Annual Review of Ecology and Systematics* 7:235–257.

Covington, W. W., and M. M. Moore. 1994. Southwestern ponderosa pine forest structure: Changes since Euro-American settlement. *Journal of Forestry* 92:39–47.

Crim, S. A. 1981. *Separate Versus Combined Resource Allocation and Scheduling: A Case Study on Two National Forests.* Ph.D. diss., Colorado State University, Fort Collins.

Cushing, J. M. 1988. Nonlinear matrix models and population dynamics. *Natural Resource Modeling* 2:539–580.

Cushing, J. M., B. Dennis, R. A. Desharnais, and R. F. Constantino. 1998. Moving toward an unstable equilibrium: Saddle nodes in population systems. *Journal of Animal Ecology* 67:298–306.

Dacey, M. 1968. A review on measures of contiguity for two and k-color maps. In B. J. L. Berry and D. F. Marble, eds., *Spatial Analysis: A Reader in Statistical Geography,* pp. 479–490. Englewood Cliffs, N.J.: Prentice Hall.

Dane, C. W., N. C. Meador, and J. B. White. 1977. Goal programming in land-use planning. *Journal of Forestry* 75:325–329.

Davis, L. S., and K. N. Johnson. 1987. *Forest Management,* 3rd ed. New York: McGraw-Hill.

Dhondt, A. A. 1988. Carrying capacity: A confusing concept. *Acta Oecologia Generalis* 9:337–346.

Diamond, J. M., and R. M. May. 1976. Island biogeography and the design of reserves. In R. M. May, ed., *Theoretical Ecology: Principles and Applications,* pp. 163–186. Philadelphia: Saunders.

Dobson, A. P., and R. M. May. 1986. Patterns of invasions by pathogens and parasites. In H. A. Mooney and J. A. Drake, eds., *Ecology of Biological Invasions of North America and Hawaii,* pp. 58–76, Ecological Studies 58. New York: Springer-Verlag.

Dolman, P. M., and W. J. Sutherland. 1997. Spatial patterns of depletion imposed by foraging vertebrates: Theory, review and meta-analysis. *Journal of Animal Ecology* 66:481–494.

Dunne, T., and L. B. Leopold. 1978. *Water in Environmental Planning.* San Francisco: W.H. Freeman.

Dunning, J. B. Jr., D. J. Stewart, B. J. Danielson, B. R. Noon, T. L. Root, R. H. Lamberson, and E. E. Stevens. 1995. Spatially explicit population models: Current forms and future uses. *Ecological Applications* 5:3–11.

Elton, C. S. 1958. *The Ecology of Invasions by Animals and Plants.* London: Methuen.

Erkut, E., C. ReVelle, and Y. Ulkusal. 1996. Integer-friendly formulations for the *r*-separation problem. *European Journal of Operational Research* 92:342–351.

Fagerstone, K. A., and D. E. Biggins. 1986. Comparison of capture–recapture and visual count indices of prairie dog densities in black-footed ferret habitat. *Great Basin Naturalist Memoirs* 8:94–98.

Fahrig, L. 1992. Relative importance of spatial and temporal scales in a patchy environment. *Theoretical Population Biology* 41:300–314.

FEMAT. 1993. *Forest Ecosystem Management: An Ecological, Economic, and Social Assessment,* Report of the Forest Ecosystem Management Assessment Team. Portland: USDA Forest Service, USDA Bureau of Land Management and USDI Fish & Wildlife Service.

Finney, M. A. 1998. *FARSITE: Fire Area Simulator—Model Development and Evaluation.* Research Paper RMRS-RP-4. Fort Collins, Colo.: USDA Forest Service.

Fisher, R. A. 1937. The wave of advance of advantageous genes. *Annals of Eugenics* 7:355–369.

Fitter, A. H., and R. K. M. Hay. 1987. *Environmental Physiology of Plants.* New York: Academic Press.

Fitzgerald, J. P. 1993. The ecology of plague in Gunnison's prairie dogs and suggestions for the recovery of black-footed ferrets. In J. L. Oldemeyer, D. E. Biggins, B. J. Miller, and R. Crete, eds., *Proceedings of the Symposium on the Management of Prairie Dog Complexes for the Reintroduction of Black-Footed Ferrets,* Biological Report No. 13, pp. 50–59. Washington, D.C.: U.S. Fish and Wildlife Service.

Forrest, S. C., T. W. Clark, L. Richardson, and T. M. Campbell III. 1985. *Black-Footed Ferret Habitat: Some Management and Reintroduction Considerations.* Wyoming Bureau of Land Management Technical Bulletin No. 2.

Forsman, E. D., S. DeStefano, M. G. Raphael, and R. J. Guitiérrez. 1996. Demographics of the northern spotted owl. *Studies in Avian Biology* 17:1– 122.

Fried, J. S., and B. D. Fried. 1996. Simulating wildfire containment with realistic tactics. *Forest Science* 42:267–281.

Fried, J. S., and J. K. Gilless. 1988. Stochastic representation of fire occurrence in a wildland fire protection planning model for California. *Forest Science* 34:948–955.

Fried, J. S., and J. K. Gilless. 1989. Expert opinion estimation of fireline production rates. *Forest Science* 35:870–877.

Game, M. 1980. Best shape for nature reserves. *Nature* 287:630–632.

Garrett, M. G., and W. L. Franklin. 1982. Prairie dog dispersal in Wind Cave National Park: Possibilities for control. In R. M. Timm and R. J. Johnson, eds., *Fifth Great Plains Wildlife Damage Control Workshop Proceedings, Institute of Agriculture and Natural Resources,* pp. 185–197. Lincoln: University of Nebraska.

Gatrell, A. C. 1977. Complexity and redundancy in binary maps. *Geographical Analysis* 9:29–41.

Geary, R. C. 1954. The contiguity ratio and statistical mapping. *The Incorporated Statistician* 5:115–145.

Giles, R. H., Jr. 1978. *Wildlife Management.* San Francisco: W.H. Freeman.

Gilless, J. K., and J. S. Fried. 1999. Stochastic representation of fire behavior in a wildland fire protection planning model for California. *Forest Science* 45:492–499.

Gilpin, M. E. 1987. Spatial structure and population vulnerability. In M. E. Soulé, ed., *Viable Populations for Conservation,* pp. 125–139. Cambridge, U.K.: Cambridge University Press.

Goodman, D. 1987. Consideration of stochastic demography in the design and management of biological reserves. *Natural Resource Modeling* 1:205– 234.

Gotelli, N. J., and G. R. Graves. 1996. *Null Models in Ecology.* Washington, D.C.: Smithsonian Institution Press.

Gove, J. H., and S. Fairweather. 1992. Optimizing the management of uneven-aged forest stands: A stochastic approach. *Forest Science* 38:623– 640.

Graham, R., P. Farnum, R. Timmis, and G. Ritchie. 1985. Using modeling as a tool to increase forest productivity and value. In R. Ballard, P. Farnum, G. Ritchie, and J. Winjum, eds., *Forest Potentials: Productivity and Value,* pp. 101–130. Tacoma, Wash.: Weyerhaeuser.

Greulich, F. E., and W. G. O'Regan. 1982. *Optimum Use of Air Tankers in Initial Attack: Selection, Basing, and Transfer Rules.* Research Paper PSW- 163. Berkeley, Calif.: USDA Forest Service.

Gurney, W. S. C., and R. M. Nisbet. 1975. The regulation of inhomogeneous populations. *Journal of Theoretical Biology* 52:441–457.

Haack, R. A., and J. W. Byler. 1993. Insects and pathogens: Regulators of forest ecosystems. *Journal of Forestry* 91:32–37.

Haeckel, E. 1866. *Generelle Morphologie der Organismen* (2 vols.). Berlin: Reimer.

Haight, R. G. 1987. Evaluating the efficiency of even-aged and uneven-aged stand management. *Forest Science* 33:116–134.

Haight, R. G., J. D. Brodie, and D. M. Adams. 1985. Optimizing the sequence of diameter distributions and selection harvests for uneven-aged stand management. *Forest Science* 31:451–462.

Haight, R. G., and R. A. Monserud. 1990a. Optimizing any-aged management of mixed-species stands: I. Performance of a coordinate-search process. *Canadian Journal of Forest Research* 20:15–25.

Haight, R. G., and R. A. Monserud. 1990b. Optimizing any-aged management of mixed-species stands: II. Effects of decision criteria. *Forest Science* 36:125–144.

Hamazaki, T. 1996. Effects of patch shape on the number of organisms. *Landscape Ecology* 11:299–306.

Hann, D. W., and B. B. Bare. 1979. *Uneven-Aged Forest Management: State of the Art (or Science?),* General Technical Report INT-50. Ogden, Utah: USDA Forest Service.

Harris, L. D. 1984. *The Fragmented Forest: Island Biogeography Theory and the Preservation of Biotic Diversity.* Chicago: University of Chicago Press.

Harris, L. D., and J. D. McElveen. 1981. Effect of forest edges on north Florida breeding birds. *IMPAC Reports* 6(4).

Harris, R. B., T. W. Clark, and M. L. Shaffer. 1989. Extinction probabilities for isolated black-footed ferret populations. In U. S. Seal, E. T. Thorne, M. A. Bogan, and S. H. Anderson, eds., *Conservation Biology and the Black- Footed Ferret,* pp. 69–82. New Haven, Conn.: Yale University Press.

Harrison, S., and A. D. Taylor. 1997. Empirical evidence for metapopulation dynamics. In I. A. Hanski and M. E. Gilpin, eds., *Metapopulation Biology: Ecology, Genetics, and Evolution,* pp. 27–42. San Diego: Academic Press.

Hassell, M. P., O. Miramontes, P. Rohani, and R. M. May. 1995. Appropriate formulations for dispersal in spatially structured models: Comments on Bascompte and Solé. *Journal of Animal Ecology* 64:662–664.

Hastings, A. 1982. Dynamics of a single species in a spatially varying environment: The stabilizing role of high dispersal rates. *Journal of Mathematical Biology* 16:49–55.

Hastings, A. 1992. Age dependent dispersal is not a simple process: Density dependence, stability, and chaos. *Theoretical Population Biology* 41:388–400.

Henderson, F. R., P. F. Springer, and R. Adrian. 1969. *The Black-Footed Ferret in South Dakota.* Technical Bulletin No. 4. Pierre: South Dakota Department of Game, Fish, and Parks.

Hengeveld, R. 1987. Scales of variation: Their distinction and ecological importance. *Annales Zoologici Fennica* 24:195–202.

Henschel, J. R., and Y. D. Lubin. 1997. A test of habitat selection at two spatial scales in a sit-and-wait predator: A web spider in the Namib Desert dunes. *Journal of Animal Ecology* 66:401–413.

Hesselyn, H., D. B. Rideout, and P. N. Omi. 1998. Using catastrophe theory to model wildfire behavior and control. *Canadian Journal of Forest Research* 28:852–862.

Hewlett, J. D., and A. R. Hibbert. 1967. Factors affecting the response of small watersheds to precipitation in humid regions. In W. E. Sopper and H. W. Lull, eds., *Forest Hydrology,* pp. 275–290. Oxford, U.K.: Pergamon.

Hillman, C. N., R. L. Linder, and R. B. Dahlgren. 1979. Prairie dog distributions in areas inhabited by black-footed ferrets. *American Midland Naturalist* 102:185–187.

Hirsch, K. G., P. N. Corey, and D. L. Martell. 1998. Using expert judgement to model initial attack fire crew effectiveness. *Forest Science* 44:539–549.

Hof, J. 1993. *Coactive Forest Management.* San Diego: Academic Press.

Hof, J., and M. Bevers. 1998. *Spatial Optimization for Managed Ecosystems.* New York: Columbia University Press.

Hof, J., M. Bevers, L. Joyce, and B. Kent. 1994. An integer programming approach for spatially and temporally optimizing wildlife populations. *Forest Science* 40: 177–191.

Hof, J. G., J. B. Pickens, and E. T. Bartlett. 1986. A MAXMIN approach to nondeclining yield timber harvest scheduling problems. *Forest Science* 32:653–666.

Hof, J., and M. G. Raphael. 1997. Optimization of habitat placement: A case study of the northern spotted owl in the Olympic Peninsula. *Ecological Applications* 7: 1160–1169.

Hof, J., D. Rideout, and D. Binkley. 1990. Carbon fixation in trees as a micro optimization process: An example of combining ecology and economics. *Ecological Economics* 2:243–256.

Hof, J., C. H. Sieg, and M. Bevers. 1999. Spatial and temporal optimization in habitat placement for a threatened plant: The case of the western prairie fringed orchid. *Ecological Modelling* 115:61–75.

Holmes, E. E., M. A. Lewis, J. E. Banks, and R. R. Veit. 1994. Partial differential equations in ecology: spatial interactions and population dynamics. *Ecology* 75:17–29.

Holthausen, R. S., M. G. Raphael, K. S. McKelvey, E. D. Forsman, E. E. Starkey, and D. E. Seaman. 1995. *The Contribution of Federal and Nonfederal Habitat to Persistence of the Northern Spotted Owl on the Olympic Peninsula, Washington,* General Technical Report PNW-GTR-352. Portland, Oreg.: USDA Forest Service.

Hoogland, J. L. 1995. *The Black-Tailed Prairie Dog: Social Life of a Burrowing Mammal.* Chicago: The University of Chicago Press.

Hoogland, J. L., D. K. Angell, J. G. Daley, and M. C. Radcliffe. 1988. Demography and population dynamics of prairie dogs. In D. W. Uresk, G. L. Schenbeck, and R. Cefkin, tech. coords., *Eighth Great Plains Wildlife Damage Control Workshop Proceedings,* General Technical Report RM-154, pp. 18–22. Rapid City, S.Dak.: USDA Forest Service.

Horton, R. E. 1935. *Surface Runoff Phenomena. Part I: Analysis of the Hydrograph.* Voorheesville, N.Y.: Horton Hydrological Laboratory Publication 101.

Houston, B. R., T. W. Clark, and S. C. Minta. 1986. Habitat suitability index model for the black-footed ferret: A method to locate transplant sites. *Great Basin Naturalist Memoirs* 8:99–114.

Hunter, M. L., Jr. 1990. *Wildlife, Forests, and Forestry: Principles of Managing Forests for Biological Diversity.* Englewood Cliffs, N.J.: Regents/Prentice Hall.

Hursh, C. R., and E. F. Brater. 1944. Separating hydrographs in surface- and subsurface-flow. *Transactions of the American Geophysical Union* 22:863–871.

Husband, B. C., and S. C. H. Barrett. 1996. A metapopulation perspective in plant population biology. *Journal of Ecology* 84:461–469.

Jager, H. I., and L. J. Gross. 2000. Spatial control: The final frontier in applied ecology. *Ecology* 81:1473–1474.

Janda, R. J., M. K. Nolan, D. R. Harden, and S. M. Colman. 1975. *Watershed Conditions in the Drainage Basin of Redwood Creek, Humboldt Co., California, as of 1973.* Menlo Park, Calif.: U.S. Geological Survey Open File Report No. 75.568.

Jarvis, P. 1985. Increasing productivity and value of temperate coniferous forest by manipulating site water balance. In R. Ballard, P. Farnum, G. Ritchie, and J. Winjum, eds., *Forest Potentials: Productivity and Value,* pp. 39–74. Tacoma, Wash.: Weyerhaeuser.

Jarvis, P., and Leverenz, J. 1983. Productivity of temperate, deciduous, and evergreen forests. In O. Lange, P. Nobel, C. Osmond, and H. Ziegler, eds., *Physiological Plant Ecology, IV, Ecosystem Processes: Mineral Cycling, Productivity and Man's Influence,* pp. 223–280. New York: Springer.

Johnson, K. N., and H. L. Scheurman. 1977. *Techniques for Prescribing Optimal Timber Harvest and Investment Under Different Objectives: Discussion and Synthesis.* Forest Science Monograph 18.

Johnson, K. N., T. W. Stuart, and S. A. Crim. 1986. *Forplan Version 2: An Overview.* Washington, D.C.: Land Management Planning Systems Section, USDA Forest Service.

Kalisz, S., and M. A. McPeek. 1993. Extinction dynamics, population growth and seedbanks. *Oecologia* 95:314–320.

Kaneko, K. 1993. *Theory and Applications of Coupled Map Lattices.* Chichester, U.K.: Wiley.

Kierstead, H., and L. B. Slobodkin. 1953. The size of water masses containing plankton blooms. *Journal of Marine Research* 12:141–147.

Knowles, C. J. 1985a. Observations on prairie dog dispersal in Montana. *Prairie Naturalist* 17:33–40.

Knowles, C. J. 1985b. Population recovery of black-tailed prairie dogs following control with zinc phosphide. *Journal of Range Management* 39:249–251.

Kot, M., and W. M. Schaffer. 1986. Discrete-time growth-dispersal models. *Mathematical Biosciences* 80:109–136.

Kourtz, P. H. 1987. Expert dispatch of forest fire control resources. *AI Applications in Natural Resource Management* 1:1–7.

Kourtz, P. H., and W. G. O'Regan. 1971. A model for a small forest fire to simulate burned and burning areas for use in a detection model. *Forest Science* 17:163–169.

Lamberson, R. H., K. S. McKelvey, and B. R. Noon. 1992. A dynamic analysis of spotted owl viability in a fragmented forest landscape. *Conservation Biology* 6:1–8.

Lamberson, R. H., B. R. Noon, C. Voss, and K. S. McKelvey. 1994. Reserve design for territorial species: the effects of patch size and spacing on the viability of the northern spotted owl. *Conservation Biology* 8:185–195.

Lande, R. 1987. Extinction thresholds in demographic models of territorial populations. *American Naturalist* 130:624–635.

Laurance, W. F., and E. Yensen. 1991. Predicting the impacts of edge effects in fragmented habitats. *Biological Conservation* 55:77–92.

Leopold, A. 1933. *Game Management.* New York: Scribner.

Levin, S. A. 1974. Dispersion and population interactions. *American Naturalist* 108:207–228.

Levin, S. A. 1989. Analysis of risk for invasions and control programs. In J. A. Drake, H. A. Mooney, F. di Castri, R. H. Groves, F. J. Kruger, M. Rejmanek, and M. Williamson, eds., *Biological Invasions: A Global Perspective, SCOPE 37,* pp. 425–435. Chichester, U.K.: Wiley.

Levin, S. A. 1992. The problem of pattern and scale in ecology. *Ecology* 73: 1943–1967.

Liebhold, A. M., J. A. Halverson, and G. A. Elmes. 1992. Gypsy moth invasion in North America: A quantitative analysis. *Journal of Biogeography* 19:513–520.

Liebhold, A. M., W. L. MacDonald, D. Bergdahl, and V. C. Mastro. 1995. Invasion by exotic pests: A threat to forest ecosystems. *Forest Science Monograph* 30.

Liebman, J., L. Ladson, L. Schrage, and A. Waren. 1986. *Modeling and Optimization with GINO.* Palo Alto, Calif.: Scientific Press.

Linder, R. L., R. B. Dahlgren, and C. N. Hillman. 1972. Black-footed ferret–prairie dog interrelationships. In *Symposium on Rare and Endangered Wildlife of the Southwestern United States,* pp. 22–37. Santa Fe: New Mexico Department of Game and Fish.

Loehle, C. 1999. Optimizing wildlife habitat mitigation with a habitat defragmentation algorithm. *Forest Ecology and Management* 120:245–251.

Long, G. E. 1979. Dispersal theory and its applications. In A. A. Berryman and L. Safranyik, eds., *Dispersal of Forest Insects: Evaluation, Theory and Management Implications,* pp. 244–250. Pullman: Washington State University Press.

Long, M. E. 1998. The vanishing prairie dog. *National Geographic* 193(4):116–131.

Luce, R. D., and H. Raiffa. 1957. *Games and Decisions: Introduction and Critical Survey.* New York: Wiley.

Ludwig, D., D. G. Aronson, and H. F. Weinberger. 1979. Spatial patterning of the spruce budworm. *Journal of Mathematical Biology* 8:217–258.

Luenberger, D. G. 1984. *Linear and Nonlinear Programming,* 2nd ed. Reading, Mass.: Addison-Wesley.

Malanson, G. P. 1996. Effects of dispersal and mortality and diversity in a forest stand model. *Ecological Modelling* 87:103–110.

Malanson, G. P., and M. P. Armstrong. 1996. Dispersal probability and forest diversity in a fragmented landscape. *Ecological Modelling* 87:91–102.

Martell, D. L. 1982. A review of operational research studies in forest fire management. *Canadian Journal of Forest Research* 12:119–140.

McDonald, P. M. 1995. *Black-Footed Ferret Monitoring, Buffalo Gap National Grassland, Winter 1994.* Unpublished report, Nebraska National Forest, Chadron.

McKelvey, K., B. R. Noon, and R. H. Lamberson. 1992. Conservation planning for species occupying fragmented landscapes: The case of the northern spotted owl. In

P. Kareiva, J. Kingsolver, and R. Huey, eds., *Biotic Interactions and Global Change*, pp. 424–450. Sunderland, Mass.: Sinauer.

McMasters, A. W. 1966. *Wildland Fire Control with Limited Suppression Forces*. Ph.D. dissertation, University of California, Berkeley.

Mees, R. M. 1985. *Simulating Initial Attack with Two Fire Containment Models*. Research Note PSW-378. Berkeley, Calif.: USDA Forest Service.

Mees, R., and D. Strauss. 1992. Allocating resources to large wildland fires: A model with stochastic production rates. *Forest Science* 38:842–853.

Mees, R., D. Strauss, and R. Chase. 1993. Modeling wildland fire containment with uncertainty in flame length and fireline width. *International Journal of Wildland Fire* 3:179–185.

Menge, B. A., and A. M. Olson. 1990. Role of scale and environmental factors in regulation of community structure. *Trends in Ecology and Evolution* 5:52–57.

Merriam, C. H. 1902. The prairie dog of the great plains. In *Yearbook of the U.S. Department of Agriculture 1901*, pp. 257–270. Washington, D.C.: U.S. Government Printing Office.

Miller, B., G. Ceballos, and R. Reading. 1994. The prairie dog and biotic diversity. *Conservation Biology* 8:677–681.

Miller, B. J., G. E. Menkens, and S. H. Anderson. 1988. A field habitat model for black-footed ferrets. In D. W. Uresk, G. L. Schenbeck, and R. Cefkin, tech. coords., *Eighth Great Plains Wildlife Damage Control Workshop Proceedings*, General Technical Report RM-154, pp. 98–102. Rapid City, S.Dak.: USDA Forest Service.

Minta, S., and T. W. Clark. 1989. Habitat suitability analysis of potential translocation sites for black-footed ferrets in north-central Montana. In T. W. Clark, D. Hinckley, and T. Rich, eds., *The Prairie Dog Ecosystem: Managing for Biological Diversity*, pp. 29–46. Billings: Montana Bureau of Land Management.

Mitchie, B. R. 1985. Uneven-aged stand management and the value of forest land. *Forest Science* 31:116–121.

Moring, J. R. 1975. *The Alsea Watershed Study: Effects of Logging on the Aquatic Resources of Three Headwater Streams of the Alsea River, Oregon, Part II— Changes in Environmental Conditions*, Fishery Research Report No. 9. Corvallis: Oregon Department of Fish and Wildlife.

Morris, D. W. 1987. Ecological scale and habitat use. *Ecology* 68:362–369.

Mulhern, D. W., and C. J. Knowles. 1997. Black-tailed prairie dog status and future conservation planning. In D. W. Uresk, G. L. Schenbeck, and J. T. O'Rourke, tech. coords., *Conserving Biodiversity on Native Rangelands Symposium Proceedings*, pp. 19–29. Fort Collins, Colo.: USDA Forest Service.

Murray, A., and R. Church. 1995. Heuristic solution approaches to operational forest planning problems. *OR Spektrum* 17:193–203.

Murray, A. T., and S. Snyder. 2000. Spatial modeling in forest management and natural resource planning. *Forest Science* 46:153–156.

Murray, J. D. 1989. *Mathematical Biology: Biomathematics*, Vol. 19. Berlin: Springer-Verlag.

Musgrave, G. W., and H. N. Holtan. 1964. Infiltration. In Ven te Chow, ed., *Handbook of Applied Hydrology* (Section 12). New York: McGraw-Hill.

Nevo, A., and L. Garcia. 1996. Spatial optimization of wildlife habitat. *Ecological Modelling* 91:271–281.

Oakleaf, B., B. Luce, and E. T. Thorne. 1992. Evaluation of black-footed ferret reintroduction in Shirley Basin, Wyoming. In B. Oakleaf, B. Luce, E. T. Thorne, and S. Torbit, eds., *1991 Annual Completion Report,* pp. 196– 240. Cheyenne: Wyoming Game and Fish Department.

Oakleaf, B., B. Luce, and E. T. Thorne. 1993. An evaluation of black-footed ferret reintroduction in Shirley Basin, Wyoming. In B. Oakleaf, B. Luce, E. T. Thorne, and B. Williams, eds., *1992 Annual Completion Report,* pp. 220–234. Cheyenne: Wyoming Game and Fish Department.

Okubo, A. 1980. *Diffusion and Ecological Problems: Mathematical Models. Biomathematics 10.* Berlin: Springer-Verlag.

Omi, P. N., J. L. Murphy, and L. C. Wensel. 1981. A linear programming model for wildland fuel management planning. *Forest Science* 27:81–94.

Orians, G. H., and N. P. Pearson. 1979. On the theory of central place foraging. In D. J. Horn, R. D. Mitchell, and G. R. Stairs, eds., *Analysis of Ecological Systems,* pp. 155–177. Columbus: Ohio State University Press.

Orians, G. H., and J. F. Wittenberger. 1991. Spatial and temporal scales in habitat selection. *American Naturalist* 137:S29–S49.

Pagel, M., and R. J. H. Payne. 1996. How migration affects estimation of the extinction threshold. *Oikos* 76:323–329.

Parks, G. M., and W. S. Jewell. 1962. *A Preliminary Model for Initial Attack,* IER Report No. 32, Operations Research Center, Institute of Engineering Research. Berkeley: University of California.

Parlar, M., and R. G. Vickson. 1982. Optimal forest fire control: An extension of Parks' model. *Forest Science* 28:345–355.

Penning de Vries, F. W. T. 1983. Modeling of growth and production. In O. Lange, P. Nobel, C. Osmond, and H. Ziegler, eds., *Physiological Plant Ecology, IV, Ecosystem Processes: Mineral Cycling, Productivity and Man's Influence,* pp. 118–150. New York: Springer.

Pielou, E. C. 1977. *Mathematical Ecology.* New York: Wiley.

Plumb, G. E., P. M. McDonald, and D. Searls. 1994. *Black-Footed Ferret Reintroduction in South Dakota: Project Description and 1994 Protocol.* Unpublished document, Badlands National Park, Wall, S.Dak.

Pukkala, T., and J. Miina. 1998. Tree-selection algorithms for optimizing thinning using a distance-dependent growth model. *Canadian Journal of Forest Research* 28:693–702.

Ramachandran, G. 1988. Probabilistic approach to fire risk evaluation. *Fire Technology* 8:204–226.

Raphael, M. G., J. A. Young, K. McKelvey, B. M. Galleher, and K.C. Peeler. 1994. A simulation analysis of population dynamics of the northern spotted owl in relation to forest management alternatives. *Final Environmental Impact Statement on Management of Habitat for Late-Successional and Old-Growth Forest Related Species Within the Range of the Northern Spotted Owl.* Volume II, Appendix J-3. Portland, Oreg.: U.S. Department of Agriculture, Forest Service; U.S. Department of the Interior, Bureau of Land Management.

Rapport, D. J., and J. E. Turner. 1977. Economic models in ecology. *Science* 195:367–373.

Rasmussen, H. N. 1995. *Terrestrial Orchids: From Seed to Mycotrophic Plant.* Cambridge, U.K.: Cambridge University Press.

Reed, W. J., and D. Errico. 1987. Techniques for assessing the effects of pest hazards on long-run timber supply. *Canadian Journal of Forest Research* 17:1455–1465.

Reeves, C. R. 1993. *Modern Heuristic Techniques for Combinatorial Problems.* New York: Wiley.

Richardson, L., T. W. Clark, S. C. Forrest, and T. M. Campbell III. 1987. Winter ecology of black-footed ferrets (*Mustela nigripes*) at Meeteetse, Wyoming. *American Midland Naturalist* 117:225–239.

Robichaud, P. R., and T. A. Waldrop. 1994. A comparison of surface runoff and sediment yields from low- and high-severity site preparation burns. *Water Resources Bulletin* 30:27–34.

Roemer, D. M., and S. C. Forrest. 1996. Prairie dog poisoning in the northern Great Plains: An analysis of programs and policies. *Environmental Management* 20: 349–359.

Roessel, B. W. P. 1950. Hydrologic problems concerning the runoff in headwater regions. *Transactions of the American Geophysical Union* 31:431– 442.

Roise, J. P. 1990. Multicriteria nonlinear programming for optimal spatial allocation of stands. *Forest Science* 36:487–501.

Rose, D. W., and C. Chen. 1977. Nonlinear biological yield models for jack pine. *Minnesota Forestry Research Note* 262.

Roughgarden, J., S. Gaines, and H. Possingham. 1988. Recruitment dynamics in complex life cycles. *Science* 241:1460–1466.

Running, S. W., and J. C. Coughlan. 1988. A general model of forest ecosystem processes for regional applications. I. Hydrologic balance, canopy gas exchange and primary production processes. *Ecological Modelling* 42:125–154.

Ruxton, G. D., J. L. Gonzalez-Andujar, and J. N. Perry. 1997. Mortality during dispersal and the stability of a metapopulation. *Journal of Theoretical Biology* 186:389–396.

Saveland, J. M., M. Stock, and D. A. Cleaves. 1988. Decision-making bias in wilderness fire management: Implications for expert system development. *AI Applications in Natural Resource Management* 2:17–29.

Schemske, D. W., B. C. Husband, M. H. Ruckelshaus, C. Goodwillie, I. M. Parker, and J. G. Bishop. 1994. Evaluating approaches to the conservation of rare and endangered plants. *Ecology* 75:584–606.

Schenbeck, G. L., and R. J. Myhre. 1986. *Aerial Photography for Assessment of Black-Tailed Prairie Dog Management on the Buffalo Gap National Grassland, South Dakota.* Fort Collins, Colo.: USDA Forest Service FPM- MAG Report No. 86-7.

Schoener, T. W. 1979. Generality of the size distance relation in models of optimal feeding. *American Naturalist* 114:902–914.

Seal, U. S. 1989. Introduction. In *Conservation Biology and the Black-Footed Ferret,* pp. xi–xvii. New Haven, Conn.: Yale University Press.

Segel, L. A., and J. L. Jackson. 1972. Dissipative structure: An explanation and an ecological example. *Journal of Theoretical Biology* 37:545–559.

Seton, E. T. 1929. *Lives of Game Animals.* Garden City, N.Y.: Doubleday.

Shelford, V. E. 1911. Physiological animal geography. *Journal of Morphology* 22: 551–618.

Shephard, R. W., and W. S. Jewell. 1961. *Operations Research in Forest Fire Prob-*

lems, IER Report No. 19, Operations Research Center, Institute of Engineering Research. Berkeley: University of California.

Sieg, C. H., and R. M. King. 1995. Influence of environmental factors and preliminary demographic analysis of a threatened orchid, *Platanthera praeclara. American Midland Naturalist* 134:61–77.

Simard, A. J. 1976. *Wildland Fire Management: The Economics of Policy Alternatives.* Ottawa: Canadian Forestry Service, Forestry Technical Report 15.

Simard, A. J. 1979. A computer simulation model of forest fire suppression with air tankers. *Canadian Journal of Forest Research* 9:390–398.

Skellam, J. G. 1951. Random dispersal in theoretical populations. *Biometrika* 38: 196–218.

Soulé, M. E. 1986. *Conservation Biology: The Science of Scarcity and Diversity.* Sunderland, Mass.: Sinauer.

Stock, M., J. Williams, and D. A. Cleaves. 1996. Estimating the risk of escape of prescribed fires: An expert system approach. *AI Applications in Natural Resource Management* 10(2):63–73.

Swersey, R. J. 1963. *Parametric and Dynamic Programming in Forest Fire Control Models,* IER Report No. 93, Operations Research Center, Institute of Engineering Research. Berkeley: University of California.

Thomas, J. W., E. D. Forsman, J. B. Lint, E. C. Meslow, B. R. Noon, and J. Verner. 1990. *A Conservation Strategy for the Northern Spotted Owl: Report of the Interagency Scientific Committee to Address the Conservation of the Northern Spotted Owl.* Portland, Oreg.: U.S. Department of Agriculture, Forest Service, U.S. Department of the Interior, Bureau of Land Management, Fish and Wildlife Service, National Park Service.

Thomas, J. W., C. Maser, and J. E. Rodiek. 1979. Edges. In J. W. Thomas, ed., *Wildlife Habitats in Managed Forests: the Blue Mountains of Oregon and Washington,* USDA Agricultural Handbook No. 553, pp. 48–59. Washington, D.C.: U.S. Department of Agriculture.

Thomas, J. W., M. G. Raphael, R. G. Anthony, E. D. Forsman, A. G. Gunderson, R. S. Holthausen, B. G. Marcot, G. H. Reeves, J. R. Sedell, and D. M. Solis. 1993. *Viability Assessments and Management Considerations for Species Associated with Late-Successional and Old-Growth Forests of the Pacific Northwest,* Report of the Scientific Analysis Team. Portland, Oreg.: USDA Forest Service.

Thorne, E. T., and D. W. Belitsky. 1989. Captive propagation and the current status of free-ranging black-footed ferrets in Wyoming. In *Conservation Biology and the Black-Footed Ferret,* pp. 223–234. New Haven, Conn.: Yale University Press.

Tiwari, S., R. Ramaswamy, and J. S. Rao. 1996. Adaptive control in a resource management model. *Ecological Modelling* 84:53–62.

Troendle, C. A. 1985. Variable source area models. In M. G. Anderson and T. P. Burt, eds., *Hydrological Forecasting,* pp. 347–403. New York: Wiley.

Turner, M. G. 1989. Landscape ecology: The effect of pattern on process. *Annual Review of Ecology and Systematics* 20:171–197.

Turner, M. G., G. J. Arthaud, R. T. Engstrom, S. J. Hejl, J. Liu, S. Loeb, and K. McKelvey. 1995. Usefulness of spatially explicit population models in land management. *Ecological Applications* 5:12–16.

Udvardy, M. D. F. 1969. *Dynamic Zoogeography.* New York: Van Nostrand Reinhold.

Uresk, D. W., J. G. MacCracken, and A. J. Bjugstad. 1981. Prairie dog density and cattle grazing relationships. In *Fifth Great Plains Wildlife Damage Control Workshop Proceedings,* pp. 199–201. Lincoln, Nebr.: USDA Forest Service.

Uresk, D. W., and G. L. Schenbeck. 1987. Effect of zinc phosphide rodenticide on prairie dog colony expansion as determined from aerial photography. *Prairie Naturalist* 19:57–61.

U.S. Department of Agriculture. 1988. *Final Supplement to the Environmental Impact Statement for an Amendment to the Pacific Northwest Regional Guide* (2 volumes). Portland, Oreg.: USDA Forest Service, Pacific Northwest Region.

U.S. Department of Agriculture. 1992. *Final Environmental Impact Statement on Management for the Northern Spotted Owl in the National Forests.* Portland, Oreg.: USDA Forest Service.

U.S. Department of Agriculture and U.S. Department of the Interior. 1994. *Final Supplemental Environmental Impact Statement on Management of Habitat for Late-Successional and Old-Growth Forest Related Species Within the Range of the Northern Spotted Owl.* Portland, Oreg.: USDA Forest Service, USDI Bureau of Land Management.

U.S. Department of the Interior. 1990. *Endangered and Threatened Wildlife: Determination of Threatened Status for the Northern Spotted Owl,* Final Rule. Portland, Oreg.: USDI Fish and Wildlife Service. Also published in *Federal Register* pp. 26114–26194, volume 55 (June 26, 1990).

U.S. Department of the Interior. 1992. *Final Draft Recovery Plan for the Northern Spotted Owl* (2 volumes). Portland, Oreg.: U.S. Fish and Wildlife Service.

U.S. Department of the Interior. 1995. *Draft Environmental Alternatives Analysis for a 4(d) Rule for the Conservation of the Northern Spotted Owl on Non-Federal Lands.* Portland, Oreg.: U.S. Fish and Wildlife Service.

U.S. Fish and Wildlife Service. 1989. Endangered and threatened wildlife and plants; determination of threatened status for *Platanthera leucophaea* (eastern prairie fringed orchid) and *Platanthera praeclara* (western prairie fringed orchid). *Federal Register* 54:39857–39862.

U.S. Fish and Wildlife Service. 1996. Platanthera praeclara *(Western Prairie Fringed Orchid) Recovery Plan.* Fort Snelling, Minn.: U.S. Fish and Wildlife Service.

U.S. Fish and Wildlife Service, National Park Service, and USDA Forest Service. 1994. *Black-Footed Ferret Reintroduction, Conata Basin/Badlands, South Dakota, Final Environmental Impact Statement.* Pierre, S.Dak.: U.S. Fish and Wildlife Service.

Valverde, T., and Silvertown, J. 1997. A metapopulation model for *Primula vulgaris,* a temperate forest understorey herb. *Journal of Ecology* 85:193– 210.

van den Bosch, F., R. Hengeveld, and J. A. J. Metz. 1992. Analysing the velocity of animal range expansion. *Journal of Biogeography* 19:135–150.

Van Deusen, P. C. 1996. Habitat and harvest scheduling using Bayesian statistical concepts. *Canadian Journal of Forest Research* 26:1375–1383.

Varley, G. C., G. R. Gradwell, and M. P. Hassell. 1973. *Insect Population Ecology.* Oxford, U.K.: Blackwell.

Von Braun, W., F. I. Ordway III, and D. Dooling. 1985. *Space Travel: A History,* 4th ed. New York: Harper & Row.

Wagner, H. M. 1975. *Principles of Operations Research,* 2nd ed. Englewood Cliffs, N.J.: Prentice Hall.

Walters, C. 1986. *Adaptive Management of Renewable Resources.* New York: Macmillan.

Wennergren, U., M. Ruckelshaus, and P. Kareiva. 1995. The promise and limitations of spatial models in conservation biology. *Oikos* 74:349–356.

Westman, W. E. 1980. Gaussian analysis: Identifying environmental factors influencing bell-shaped species distributions. *Ecology* 61:733–739.

With, K. A., and T. O. Crist. 1995. Critical thresholds in species' responses to landscape structure. *Ecology* 76:2446–2459.

Wu, J., and S. A. Levin. 1997. A patch-based spatial modeling approach: Conceptual framework and simulation scheme. *Ecological Modelling* 101:325–346.

Wuerthner, G. 1997. Viewpoint: The black-tailed prairie dog-headed for extinction? *Journal of Range Management* 50:459–466.

Yoshimoto, A., R. G. Haight, and J. D. Brodie. 1990. A comparison of the pattern search algorithm and the modified PATH algorithm for optimizing an individual tree model. *Forest Science* 36:394–412.

Zimmerman, T. Z., and D. L. Bunnell. 1998. *Wildland and Prescribed Fire Management Policy: Implementation Procedures and Reference Guide.* Boise, Idaho: National Interagency Fire Center.

INDEX

adjacency relationships: and proximity relationships, 11; and sedimentation, 16, 22; and stormflow, 41
Adler, F. R., 96
Agriculture, U.S. Department of (USDA), 60, 104
Allee effects, 96, 97
Allen, L. J. S., 74, 85, 87, 201, 204
Anderson, D. H., 184
Anderson, D. J., 49, 51
animal species: livestock, 107, 114, 115, 116; modeling of, 5; optimizing behavior of, 201; and proximity relationships, 3–4; threatened, 126. *See also* ferret, black-footed; owl, northern spotted; prairie dog, black-tailed
Apa, A. D., 107
archipelago systems, 89, 95. *See also* island systems
attractors, 202, 206

Badlands National Park (South Dakota): ferrets in, 99, 103–6, 108, 112; prairie dogs in, 120, 122, 123
Bare, B. B., 43, 49, 51
barrier zones, 168, 171, 174, 175
Bazzaz, F. A., 221
Bettinger, P., 22
Bevers, M., 78, 144, 164, 171, 204
Biggins, D. E., 99, 106
biodiffusion theory, 99, 201
biomass, 222, 223, 230, 231
Bloom, A. J., 221, 222, 223, 224, 226, 231

Boston, K., 234
branch-and-bound algorithms, 47, 55, 160
breeding. *See* reproduction
Buffalo Gap National Grassland (South Dakota): ferrets in, 99, 102–4, 106–9, 112; prairie dogs in, 115, 120, 122, 123
burning: controlled, 44; prescribed, 24, 161, 196. *See also* fire
Byler, J. W., 167

carbohydrate allocation, 222, 227, 230, 231
carbon fixation, 4, 202, 221–32
carrying capacity: and exotic pest models, 169; of ferret habitat, 100, 102, 103, 106–10, 112; and habitat edges, 147, 149; and intrapatch heterogeneity, 83–85; in island systems, 89, 91; as limiting factor, 203, 206, 217, 220; in optimization models, 163, 164–65; of orchid habitat, 127–32, 135; of owl habitat, 66–67, 68; of prairie dog habitat, 115, 116, 118–20; and reaction-diffusion models, 77, 80, 81, 82
catastrophe theory, 183
Chen, C., 16, 30
Chiang, A. C., 1
Cincotta, R. P., 107, 116, 117
climate, sensitivity to, 4, 125–31, 133, 134, 135, 138, 140, 233
clustering, population, 217–20, 233
community-level models, 234

Conata Basin (Buffalo Gap National Grassland, South Dakota), 120, 121, 122, 123
control models, 3, 4, 163–200
Coughlan, J. C., 227, 230

Diamond, J. M., 92
diffusion: bio-, 99, 201; Fick's law of, 168; models of population, 73. *See also* dispersal; reaction-diffusion models
disease, 98, 114, 115, 123. *See also* pests, exotic
dispersal: biased, 102; constant periodic, 233; of exotic pests, 168, 169, 170, 172, 179, 182; of ferrets, 99, 104, 105, 108, 112, 115; and habitat edges, 143, 145–48, 150, 151, 159; and heterogeneous habitat, 214; linear model of, 102; and multiscale limiting factors, 205, 208, 210–13, 217, 218; in optimization models, 163, 164, 165; of orchid seeds, 12, 124, 125, 127, 128, 131–39; of owls, 71; population-dependent, 4; of prairie dogs, 4, 115, 116, 118–19, 120, 122; and proximity relationships, 12; and reaction-diffusion models, 73–77, 78, 96–97, 101, 102; saturation-dependent, 114–25, 233; and single-patch habitat, 95; and topography, 124, 125–41
Dunne, T., 29

ecological models, 1, 3, 8, 142; and economic models, 231
economic models, 221–23, 231
ecosystems, 3; detail in, 234; and multi-scaled limiting factors, 203–20; optimization in, 201–32; in optimization models, 221, 223–24, 233
ecotones, 143, 161
edge effects. *See* habitat edges: effects of
equilibrium, population, 81, 83–84, 86–87; and heterogeneous habitat, 214, 215, 216; and island systems,

89, 90, 91; and multiscale limiting factors, 203, 206, 207, 208–16, 217, 218; and optimization models, 201, 202, 221; and pests, 167; and single-patch habitat, 94, 95
expert systems, 184
extinction, 95, 105, 125, 126, 154
extinction thresholds: and habitat, 213, 233; and multiscale limiting factors, 204, 211, 212, 217

federal lands, 60, 69, 70
ferret, black-footed (*Mustela nigripes*), 4, 98–113; model for, 99–106; and prairie dogs, 98–99, 106–7, 108, 112, 115, 119, 123–24; and reaction-diffusion models, 115; release of captive-bred, 98–99, 100, 102, 103–4, 106, 108, 109, 111, 164; results for, 107–13
Fick's law of diffusion, 168
fire, 4, 183–200; and binding burn path, 189–96, 197; case example of, 187–89; and fuel management, 184, 196, 199, 200; model of, 185–87; prediction models for, 184, 187, 200; resources for management of, 187; results for, 189–96; suppression of, 184; timing ignition of, 165, 184, 193, 194, 195, 196, 199; timing spread of, 184, 196
Fish and Wildlife Service, U.S., 60
Flather, C. H., 78, 204
foraging, 203, 219; and habitat edges, 143, 144, 145, 146, 147, 148, 149–55, 156, 160
forest management, 13; any-aged, 42–58; even-aged, 42, 43, 51, 52; and habitat edges, 142, 147, 148, 155, 156, 157, 159; and linear programming, 5–8; partial retention, 42; and stormflow, 24, 25, 28, 29, 41; uneven-aged, 42, 43, 48, 49, 53, 56, 57. *See also* burning; timber harvesting
forests: clearcutting of, 16, 22, 42, 43; and exotic pests, 167; late-succes-

sional and old-growth, 60–61; productivity models for, 222, 227; and stormflow, 28, 29, 30, 32–33, 41

Forest Service (USDA), 104

Franklin, W. L., 119

Fried, B. D., 184

Fried, J. S., 184

Game, M., 83, 92

Game Management (Leopold), 142

Garrett, M. G., 119

geographic information systems (GIS), 16, 17, 30

Gilless, J. K., 184

Grace, J., 221

Gross, L. J., 234

gypsy moth, 174

Haack, R. A., 167

habitat: cellular, 78, 99; connectivity of, 12, 59, 61, 65–66, 68, 69, 71, 75; contiguity of, 213, 215, 218; conversions of, 126; ephemeral, 125; experiments with, 79–80; extinction thresholds in, 233; ferret, 98–113; fragmented, 73, 74, 79–80, 85–89, 94, 95, 96, 98, 104, 234; heterogeneous, 80, 83–85, 94, 204, 213–16; homogeneous, 211; interior of, 142, 143, 149, 150, 151, 152, 155; as limiting factor, 211; loss of, 114; models of, 61–71; multiple-patch, 85–92; multiscaled, 203; orchid, 125–41; owl, 59–71; and patch shape, 83, 85, 92–93, 96; and patch size, 96, 99; patchy, 126; placement of, 73; and population, 78–97, 120–23; prairie dog, 113, 115–16; preferred, 108, 217–18; protection of, 136, 137, 138, 140, 141; single-patch, 79, 80–85, 93, 94; stepping-stone, 91, 92, 96; unfragmented circular, 161

habitat edges, 4; and adjacency relationships, 11; and breeding territories, 159, 161; case example of, 147–48; definition of, 143, 159; effects of, 142–61; model for, 143–47; positive *vs.* negative effects of, 151, 152, 160, 161, 233; and resources, 142–43; results for, 148–61

Haight, R. G., 43

Hamazaki, T., 92

Hann, D. W., 43

Harris, R. B., 105

Harrison, S., 95

Henderson, F. R., 98

heterogeneity: habitat, 94, 204, 213–16; intrapatch, 80, 83–85

Hillman, C. N., 106

Hirsch, K. G., 184

Hof, J., 164, 171, 222

Holmes, E. E., 93

Holthausen, R. S., 60, 65, 71

Hoogland, J. L., 119

Houston, B. R., 99

Hunter, M. L., Jr., 142, 143, 149

integer programming, 2, 5, 11–12, 24, 44, 47, 234

Interior, U.S. Department of the, 60

island systems, 89–92, 95

Jackson, J. L., 73

Jager, H. I., 234

Jarvis, P., 227, 230

Jewell, W. S., 183

Karush-Kuhn-Tucker conditions, 225, 226, 228

Kierstead, H., 73

Knowles, C. J., 107

Kot, M., 77

Lambert-Beer equation, 226–27

larch casebearer (*Coleophora laricella*), 169

Leopold, Aldo, 142

Leopold, L. B., 29

Leverentz, J., 227, 230

Levin, S. A., 74

Liebhold, A. M., 167, 168, 169, 172, 174

life cycles, 4, 203

limiting factors: breeding site capacities as, 203, 205, 206, 217, 219; multiple, 4; in plants, 222, 224, 226

limiting factors, multiscale, 203–20; case example of, 205–8; discussion of, 216–20; and dispersal, 205, 208, 210–13, 217, 218; and extinction thresholds, 204, 211, 212, 217; and population equilibrium, 203, 206, 207, 208–16, 217, 218; and reaction-diffusion processes, 209, 210, 212, 217, 218, 219; results for, 208–16

linear programming, 2–3, 233–34; and exotic pests, 172; and habitat edges, 143–47; at multiple spatial scales, 202; of multiscale limiting factors, 204, 206; and proximity relation-ships, 12; and reaction-diffusion models, 3–4; of stormflow, 24–41; and timber harvesting, 13; tradi-tional, 5–8

livestock, 107, 114, 115, 116

Long, G. E., 169

Ludwig, D., 83

McMasters, A. W., 183

MacMillan Bloedel company, 42

management: adaptive, 3, 140, 182, 234; and adjacency relationships, 11; of ecosystems, 1–2; and fire control, 187, 196, 199; and habitat conver-sion, 126; of habitat edges, 142, 144, 150, 151, 159–61; and limited resources, 2, 187; monitoring of, 234; multiple-resource, 2; optimal, 150, 151, 152; in optimization models, 163, 164; of orchid habitat, 135, 136, 138, 140, 141; and pest control, 168, 169–82; of prairie dogs, 103, 104, 112; of prairie dogs *vs.* ferrets, 123–24; and sedimentation, 22, 23; short *vs.* long-term, 103; spatial, 168, 171, 179; spatially dispersed strategy of, 182; storm-flow, 3, 24–41. *See also* forest management

Manitoba (Canada), 126

marine organisms, 203

Martell, D. L., 183

MAXMIN models: in fire control, 193, 195, 196, 199; in forest management, 49–51, 226, 227, 229, 231; in habitat edge management, 144, 158

May, R. M., 92

Mees, R. M., 184

Mellette County (South Dakota), 106

Merriam, C. H., 114

metapopulations, 90–91, 95, 96, 126, 201

Michelis-Menten mass action kinetics, 227

Miller, B. J., 99

minimax models, 22, 36–37, 39, 207

mixed-integer programming, 159, 161

Monserud, R. A., 43

Moring, J. R., 17

mortality: dispersal, 80, 82, 86, 88, 96, 102, 116, 119, 146, 151; pest, 171, 172, 179; winter, 106

Myhre, R. J., 104

National Park Service, 104. *See also* Badlands National Park

Needle Branch watershed, 16

NFP (Northwest Forest Plan), 60

nonlinear programming, 2–3, 5, 183, 202, 228, 234; and activity analysis, 8–10

North Dakota, 126, 127, 134, 136

Northwest Forest Plan (NFP), 60

Nuernberger, B., 96

Oakleaf, B., 105

Olympic Peninsula (Washington), 60–61, 70

optimal control theory, 183

optimization: and carbon fixation, 221–32; constrained vs. uncon-strained, 1; defined, 1; in ecosystems, 201–32; of management, 150, 151, 152; and multiscale limiting factors, 203–20; natural selection as, 201; and proximity relationships, 3–4, 60, 68

optimization models, spatial: linear, 8,

117, 233, 234; nonlinear, 234; prescriptive, 201; process-oriented, 234; timing-variable, 200. *See also* under particular topics

orchid, western prairie fringed (Platanthera praeclara), 125–41; dispersal of, 124, 125, 128, 131–39; life stages of, 127; model for, 128–36; population dynamics of, 126–28, 130–31, 135, 137, 138, 139, 140; results for, 136–41; and topography, 128, 131, 134–36

Orians, G. H., 143

owl, northern spotted (*Strix occidentalis caurina*), 59–71, 234; model for, 61–67; results for, 67–71

Parks, G. M., 183

pests, exotic, 164, 167–82; and adjacency relationships, 11; barrier zones for, 168, 171, 174, 175; case example of, 171–82; establishment of, 167, 168; and landscape features, 174, 179, 182; results for, 172–82

plague (*Yersinia pestis*), 98, 114, 115, 123

plant species, 4, 5; and economic models, 221–22; in optimization models, 223–24; optimizing behavior of, 201; productivity of, 222; spatial patchiness of, 125; threatened, 126. *See also* orchid, western prairie fringed

population: abundance of, 203, 204, 206, 216–17; age-structured, 75; and carrying capacity, 108; density of, 74, 96, 97; dispersal of, 61, 89–92; distribution of, 147–61, 203, 204, 206, 216–17; fragmented, 204, 212, 217, 219; and habitat fragmentation, 73; and habitat structure, 78–97; maximization vs. minimization of, 4, 163–64, 170; multiscale limiting factors on, 203–20; and optimization, 201; orchid, 126–28, 130–31, 135, 137–40; owl, 67–70; plant, 125; prairie dog, 115–17, 120–23; rates of

change of, 74; and reaction-diffusion models, 96; recovery of, 92; and simulation models, 59

population growth, 61; density-dependent, 96, 97; ferret, 102–3, 105–6, 108; and habitat increase, 120–23; and island systems, 90–91, 92; and multiscale limiting factors, 211, 218; in optimization models, 163, 164; and patch shape, 83, 93, 94; and patch size, 82; and reaction-diffusion models, 101, 102; sigmoid curves of, 108, 233; and single-patch habitat, 93, 94

prairie dog, black-tailed (*Cynomys ludovicianus*), 4, 114–24; and ferrets, 98–99, 106–7, 108, 112, 115, 119, 123–24; habitat of, 113, 114–15; management of, 103, 104; model for, 115–20; and population control, 102; results for, 120–24

precipitation, 3, 124, 127, 135; and stormflow, 28, 29, 31

predation, 203, 205, 217, 219

propagation waves of invasion, 98–113, 167–82, 233

proximity relationships: and adjacency relationships, 11; and natural regeneration, 12, 42–58; and optimization, 3–4; and optimization models, 60, 68; and sedimentation, 13–23; simple, 11–71; and simulation models, 59–71; and stormflow management, 24–41

Public Land Survey, U.S., 104

quasirandom distributions, 203–20, 233

random walk model, 73, 74, 80, 204

Raphael, M. G., 60

Rapport, D. J., 221

reaction-diffusion models, 5, 73–161, 201; and carrying capacity, 77, 80, 81, 82; and control models, 164; discrete, 78–97; and dispersal, 73–77, 78, 96–97, 101, 102; and exotic pests, 168; and ferrets, 115;

reaction-diffusion models (*continued*), and habitat edges, 146, 150, 153, 154, 160, 161; and heterogeneous habitat, 213, 215; and linear programming, 3–4; multipatch, 74, 218; and multiscale limiting factors, 204, 209, 210, 212, 217, 218, 219; population-dependent, 115; and population growth, 101, 102; and reproduction, 74, 75, 77; single-patch vs. multipatch, 74; threshold effect in, 80, 81, 82

Reeves, C. R., 2

regeneration, 3; artificial, 16, 30, 43, 46; and incomplete initial seeding, 54–58; natural, 42–58; and proximity relationships, 12; and sedimentation, 13

release programs, captive-bred ferret, 98–100, 102–4, 106, 108, 109, 111, 164

reproduction: and breeding seasons, 77; and breeding sites, 203, 205, 206, 213, 217, 219; and breeding territories, 74, 159, 161; constant periodic, 233; of exotic pests, 172, 179; ferret, 100, 102, 103, 112; and habitat, 78, 203; and habitat edges, 143–49, 151, 152, 154, 156, 159, 160; and island systems, 90–91, 95; and multiscale limiting factors, 204, 205, 206, 208, 210–13, 217, 218; prairie dog, 116–17; and reaction-diffusion models, 74, 75, 77; and single-patch systems, 96

resources: allocation of, 2, 59, 183–84, 202; in economic models, 221; and fire control, 183–84, 187; in forest productivity models, 222; plant, 223; scarce, 201, 221; of trees, 202

rodenticide control programs, 99, 100, 102, 107, 108, 112, 114–16, 120, 123

Rose, D. W., 16, 30

Running, S. W., 227, 230

Sage Creek Wilderness (Badlands National Park, South Dakota), 103, 104

Saveland, J. M., 184

Schaffer, W. M., 77

scheduling models: dynamic, 155–59; forest, 25

Schenbeck, G. L., 104

Seal, U. S., 98

sedimentation, 3, 13–23; case example of, 15–18; and clearcutting, 16, 22; formulation of, 14–15; results for, 18–23; and stormflow, 24, 37

seed banks, 127, 138

seeding, incomplete initial, 54–58

seeds: dispersal of, 12, 124, 125, 127, 128, 131–39; dormancy of, 125; production of, 128; and proximity relationships, 12; viability of, 125, 128, 129, 130

Segal, L. A., 73

Shephard, R. W., 183

Sheyenne National Grassland (North Dakota), 126, 127, 134, 136

Shirley Basin (Wyoming), 99, 105

simulation models, 16, 221; discrete-event Monte Carlo, 234; and fire control, 183; and linear programming, 233–34; and multiscale limiting factors, 204; and optimization models, 59–71, 201; of owl habitat, 59–71; and spatial optimization models, 59–71; of stormflow, 25, 28, 32

Skellam, J. G., 73, 80, 169, 210

Slobodkin, L. B., 73

South Dakota, 106, 115. *See also* Badlands National Park; Buffalo Gap National Grassland

Spatial Optimization for Managed Ecosystems (Hof and Bevers), 2

species-habitat association, 218

stochastic effects, 84, 102, 112, 234

Stock, M., 184

stormflow, 3, 24–41, 199; case example of, 30–32; forumulation of, 25–30; hydrographs of, 233; results for, 32–41; and soil, 24, 29

Swersey, R. J., 183

Taylor, A. D., 95
threatened species, federal list of, 126
threshold effects, 87; in island systems, 89, 90; and patch size, 93; in reaction-diffusion models, 80, 81, 82; and single-patch habitat, 95. *See also* extinction thresholds
timber harvesting, 3; and adjacency relationships, 11; and habitat edges, 155, 156, 159; and natural regeneration, 42–58; revenue from, 28, 34, 35, 46, 49, 50, 54, 58; and sedimentation, 13–23; and stormflow, 24, 25, 30, 32, 34–41
topography: and dispersal, 124, 125–41; and fire control, 165, 187, 188, 192; of orchid habitat, 128, 131, 134–36
trees, 3; allocation of resources by, 202; carbon fixation in, 4, 202, 221–32; as choice variables, 43, 47. *See also* forests; timber harvesting
Turner, J. E., 221
Turner, M. G., 1

Uresk, D. W., 104

USDA (U.S. Department of Agriculture), 60, 104

Varley, G. C., 169
vegetative manipulation, 3, 144. *See also* forest management; plant species

water: and adjacency relationships, 11; ground, 135, 140; and proximity relationships, 12; quality of, 22, 41. See also stormflow
watersheds, 3; and proximity relationships, 12; and sedimentation, 13–23; and stormflow, 25, 30, 33, 37, 41
water table, 126
Wennergren, U., 96
wetlands, 125, 126
Weyerhauser, 42
wildfire. See fire
Wittenberger, J. F., 143
Wyoming Game and Fish Department, 98